IMAGES
of America

FIREFIGHTING IN FREDERICK COUNTY

IMAGES
of America

FIREFIGHTING IN FREDERICK COUNTY

Clarence "Chip" Jewell

ISBN 0-7385-4207-5

Published by Arcadia Publishing
Charleston SC, Chicago IL, Portsmouth NH, San Francisco CA

Printed in Great Britain

Library of Congress Catalog Card Number: 2005931164

For all general information contact Arcadia Publishing at:
Telephone 843-853-2070
Fax 843-853-0044
E-mail sales@arcadiapublishing.com
For customer service and orders:
Toll-Free 1-888-313-2665

Visit us on the Internet at www.arcadiapublishing.com

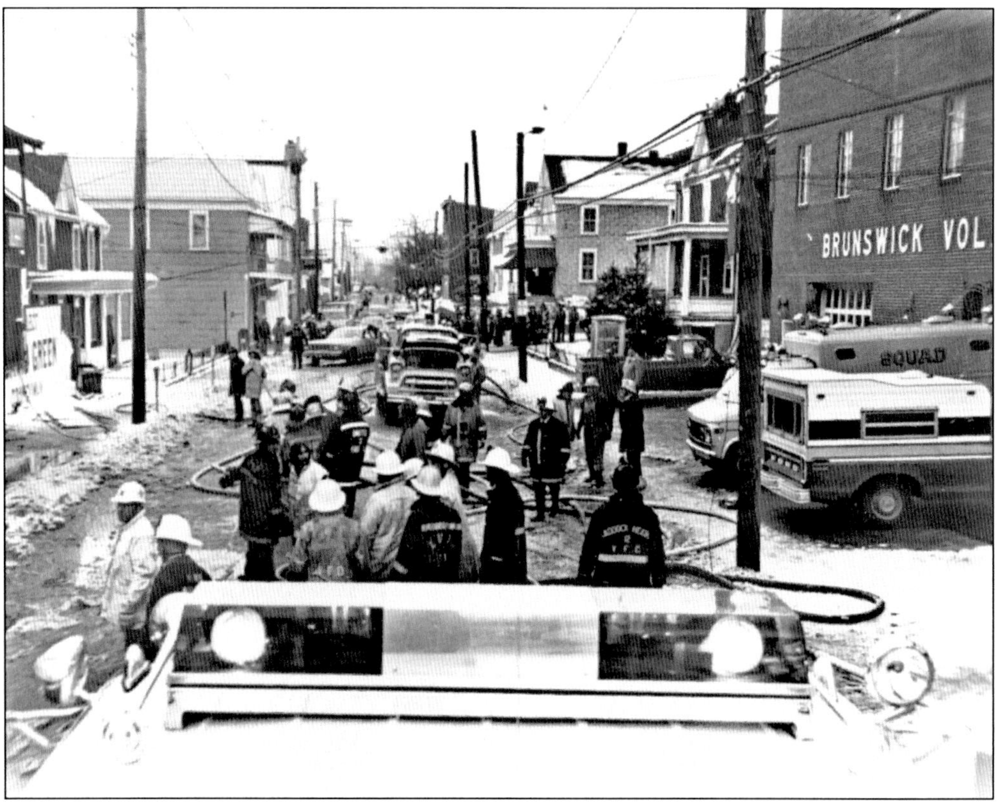

Looking over the lights of an engine, firefighters battle a winter blaze at the shoe store across from the Brunswick Volunteer Fire Company. (Courtesy Brunswick Volunteer Fire Company.)

Contents

ACKNOWLEDGMENTS

I would like to express my sincere appreciation to the many members of the fire and rescue companies throughout Frederick County for offering their pictures to make this book a success. However, as with any major project, there are a few people who went the extra mile and put in the extra effort to assist in this endeavor.

Lt. William Metcalf, Division of Fire/Rescue Services assigned to the Junior Fire Company No. 2 and a life member of that company, took on the task of scanning many of the pictures you will see as you read this book. His expertise and extra dedication of personal time was instrumental in the publication of this book.

Mr. John Kinnaird, member of the Thurmont Historical Society and of the Guardian Hose Company, took the extra time to provide some unique historical pictures for use in the book, including the cover picture.

Mr. Dave Snyder of the Jefferson Volunteer Fire Company made a concerted effort to provide pictures on CD ready for publication. He also provided pictures that could be used in various areas of the book.

Mr. Jim Wiegand, a member of the Braddock Heights Volunteer Fire Company and a former dispatcher with the Emergency Communications Center, provided pictures from his entire collection, which were used in several chapters.

My copartner in the first book, *Firefighting in Frederick*, Mr. Warren Jenkins, provided many of the pictures of the more recent apparatus used in this book.

Susie Nicol Kyle, life member of the Brunswick Volunteer Ambulance Company, provided a variety of pictures from her personal collection.

All profits from this book will be donated to the Maryland Fire/Rescue Memorial Foundation and the National Fallen Firefighters Foundation.

INTRODUCTION

As the Monocacy Trail from Lancaster County, Pennsylvania, became an avenue for western settlement, the trail also became the front door for settlement of Frederick County, Maryland. As villages were settled in the early 18th century, Frederick County became a critical area for agriculture and development to the west. The German settlers arrived in an area south of the Monocacy River, where Jacob Brunner built a house named after his beloved German community, Schifferstadt. In 1745, Frederick became a growing Maryland city.

Today the very families that helped settle our area still have direct descendants who are active in the Frederick County fire and rescue service. Charles Brunner serves as president of the Walkersville Volunteer Fire Company, and he and his brother have served as volunteers with the United Fire Company No. 3. Both men also retired as career firefighters in the Montgomery County Fire and Rescue Service.

Though the fire service of Frederick dates back to 1760, the Independent Hose Company No. 1 was not organized until 1818. However, the predecessor company, the Frederick Hose Company, dates to the purchase of the first fire engine in 1764. Many other communities in Frederick County had fire protection around 1800.

In 1803, the town of Middletown purchased a fire engine for $400. Money was raised by means of a lottery. Two thousand tickets were sold at $1 each with $1,600 distributed as prizes. Around 1850, the ladies of the town held a fair and raised money to help purchase a new engine. Approximately $900 was earned, and a new hand pumper was purchased.

Once this engine was received, a fire company was organized with George Bowlus serving as president. The newly formed company took the name Independents, possibly to mirror its neighbors in Frederick City. Oddly a rival company took possession of the original engine and called themselves the Juniors, another possible reference to the Frederick fire companies. Both companies were fully equipped with ladders, hooks, and buckets. These items were strategically placed throughout Middletown to be handy in the event of an alarm.

The Middletown Independents boasted 54 members, while the competing Juniors listed about 40 strong. Each fireman was supplied with a leather bucket to respond to an alarm of fire. At times, competition was fierce, and a number of scrimmages occurred between the two groups as they vied to extinguish the fires. The original companies were disbanded sometime around the outbreak of the Civil War. In 1894, the town burgess and commissioners took the lead to organize a fire company. The Middletown Volunteer Fire Company of today was not organized until March 6, 1923.

The fire protection in Emmitsburg, the town that is home to the National Fire Academy, can be traced to 1757, when every able-bodied person was expected to participate as a member of bucket brigades. A notice to John Bader dated May 25, 1829, requested Mr. Bader to "attend at the engine house . . . with buckets and other requisites for company training." In 1840, an outdated suction pumper was purchased to be able to pump water to fight a fire. The present Vigilant Hose Company was organized in 1884 and has been in continuous operation since that date.

In 1887, a fire company was organized in neighboring Mechanicstown. The early Guardian Hose Company depended on hand-drawn equipment to protect the young town. The company also organized one of the first drill teams and maintained this precision team for many years. In 1893, the town name was changed to Thurmont, and in 1890, a town hall, equipped with a fire bell, was built with accommodations for the Guardian Hose Company to keep the equipment.

Many communities suffered major fires. Emmitsburg, Myersville, Middletown, and Libertytown all suffered disastrous fires. The organization of formal fire protection in these communities is a direct result of these major fires. Fires that were possibly intentionally set during the Civil War may have also contributed to the need for fire protection in several towns around Frederick County.

After 1900, many towns made fire protection a priority. Towns that had fire companies that faltered prior to 1900 reorganized and incorporated many of the fire companies in existence today. Motorized equipment was purchased in the larger towns, and several of the smaller towns took possession of the unused apparatus to help provide fire protection in the more rural areas.

As the county continued to grow, additional service demands were placed on the departments. Up to 1950, ambulance service was only available from local funeral homes and the Maryland State Police. Fire companies, American Legion and VFW posts, and newly organized volunteer ambulance companies began to supplement existing services, eventually providing primary emergency medical services for the county.

Frederick County has grown from just seven organized fire companies in 1899 to 26 separate volunteer fire and rescue companies responding from 30 stations throughout Frederick County in 2005. The sleepy agricultural county is now a growing metropolitan county that is a suburb for both Washington and Baltimore. Some companies respond to as many as 5,000 emergency calls per year.

In *Firefighting in Frederick County*, the very early organizations, as well as today's modern equipment, are featured to provide the reader with a historical and practical perspective of the fire and rescue service in Frederick County, Maryland. This book complements the information in *Firefighting in Frederick*, another Arcadia Publishing book that concentrated on the fire companies located in the city of Frederick.

One

The Early Years
1800 to 1900

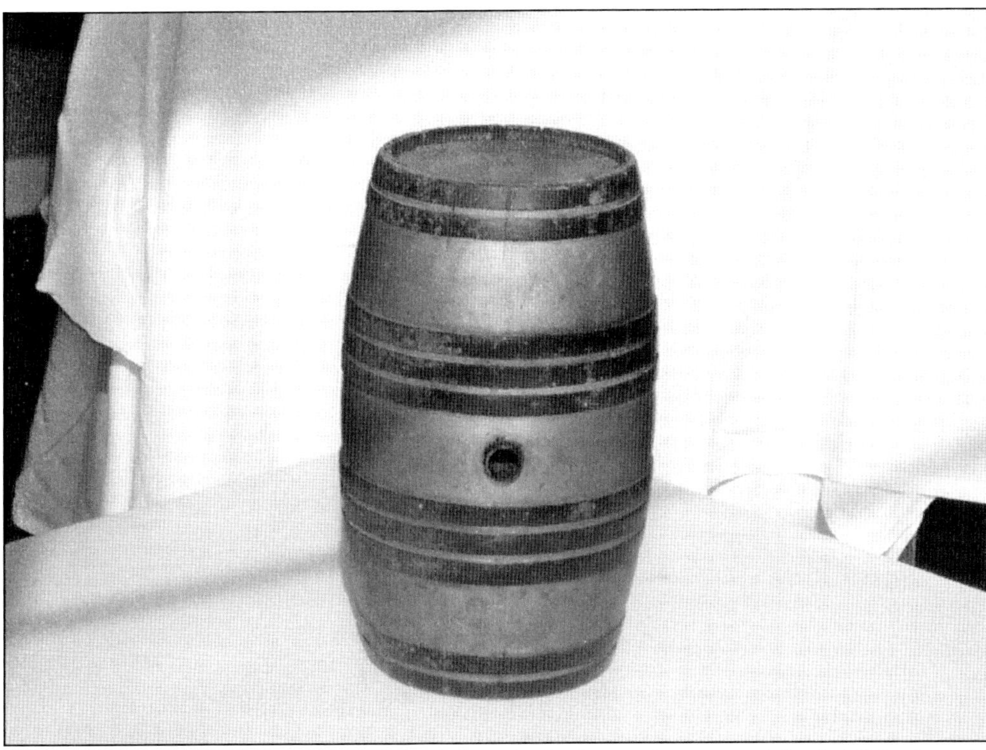

Volunteer firefighters are sometimes a lively bunch. Such was the case on May 1, 1896, when the lads did a commendable job extinguishing a fire at the George Hauser Brewery. However, a "battle royal" broke out between members of responding companies to claim the beer barrel perched atop the sign. The Independent Hose Company No. 1 won the altercation, and the beer barrel is proudly displayed to this day in the trophy case at the firehall. (Courtesy Richard Brandenburg, Independent Hose Company [IHC].)

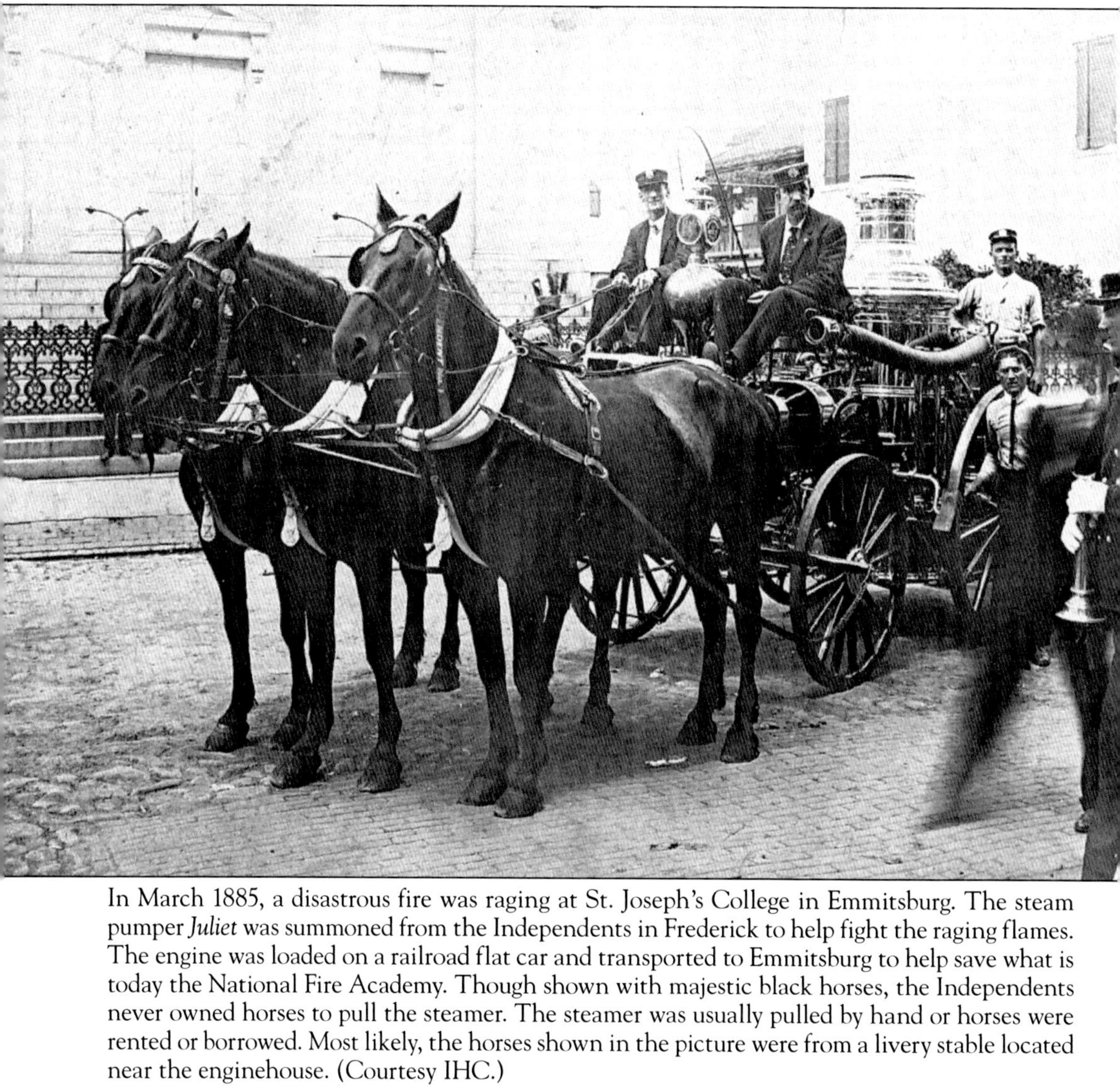

In March 1885, a disastrous fire was raging at St. Joseph's College in Emmitsburg. The steam pumper *Juliet* was summoned from the Independents in Frederick to help fight the raging flames. The engine was loaded on a railroad flat car and transported to Emmitsburg to help save what is today the National Fire Academy. Though shown with majestic black horses, the Independents never owned horses to pull the steamer. The steamer was usually pulled by hand or horses were rented or borrowed. Most likely, the horses shown in the picture were from a livery stable located near the enginehouse. (Courtesy IHC.)

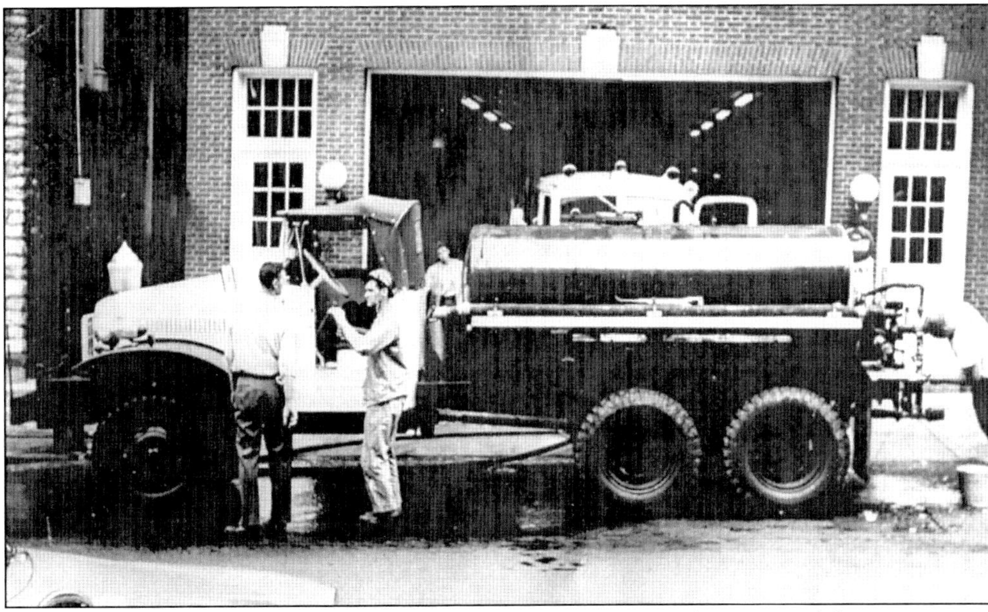

The Independent Hose Company No. 1 is the oldest continuously organized fire company in the state of Maryland. Though located in Frederick City, the Independents regularly responded to emergencies outside the city limits. In fact, for many years, they were the only city company to respond first due (the fire company to respond first) in the county areas. Thus, they were the first company to purchase a tanker in Frederick County. Shown is the army-surplus tanker that responded with additional water for calls in non-hydranted areas of Frederick County. Shown from left to right are Jack Fleischman, Bob Rinehart, and Mike Staley, with an unidentified member in the background. (Courtesy IHC.)

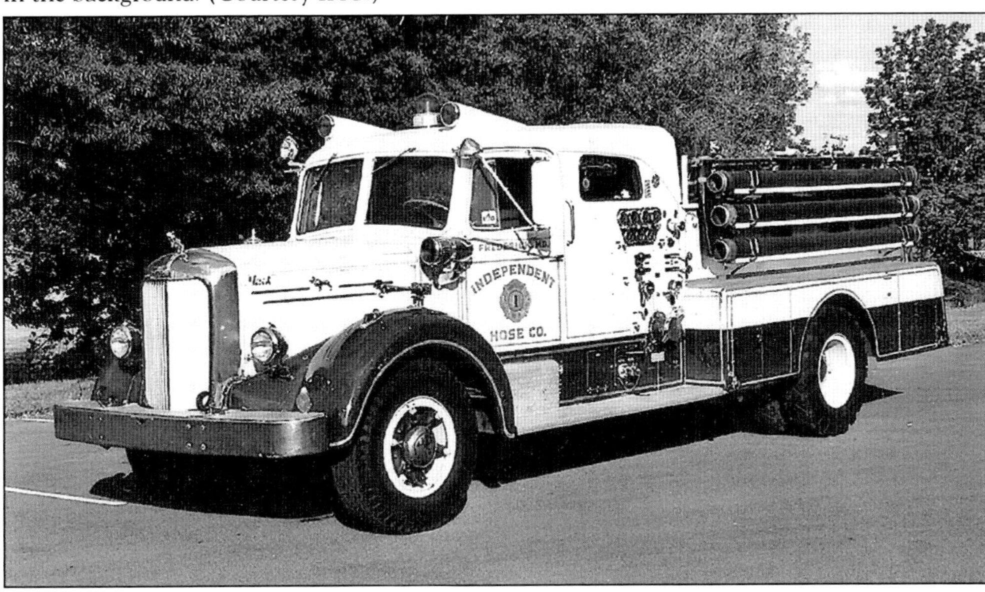

The classic 1953 Mack was the "county engine" that carried 400 gallons of water and had an 11-man cab to transport personnel. This engine remained in service until the early 1970s, responding to calls throughout Frederick County. If assistance was needed, the sister "city engine" was dispatched. (Courtesy Warren Jenkins.)

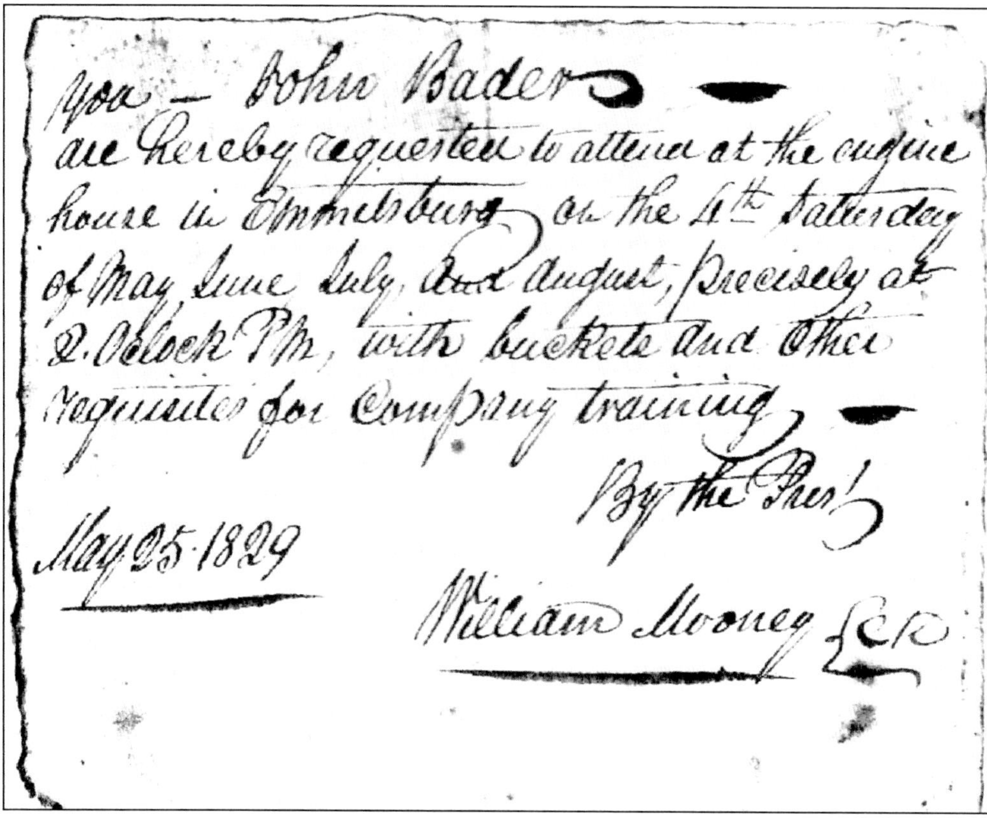

You — John Bader —
are hereby requested to attend at the engine
house in Emmitsburg on the 4th Saturday
of May June July and August, precisely at
2 O'clock PM, with buckets and other
requisites for Company training —
By the Pres!
May 25 1829
William Mooney

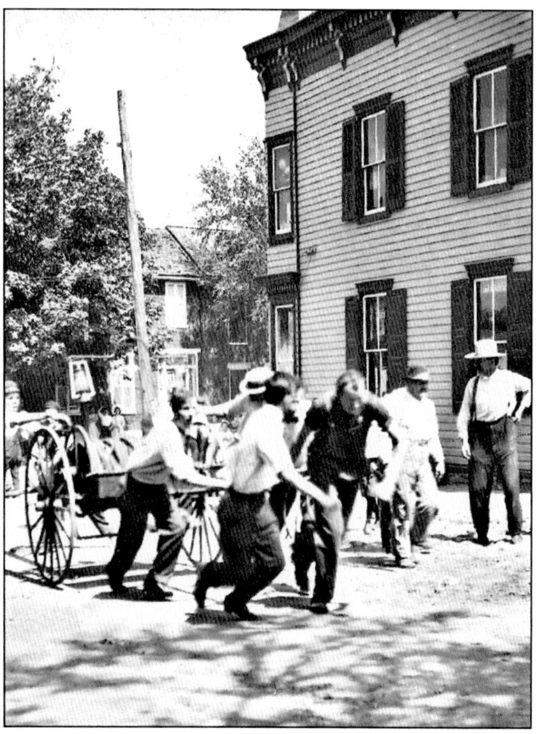

Many communities had organized fire protection before the formal fire companies existed. Such was the case in Emmitsburg, as this notice dated May, 25, 1829 states: "You—John Bader—are hereby requested to attend at the engine house in Emmitsburg on the 4th Saturday of May, June, July and August, precisely at 2 O'Clock PM, with buckets and other requisites for Company training—by the Pres. William Mooney." (Courtesy Vigilant Hose Company [VHC].)

Members of the community pull a hose reel through the streets of Emmitsburg to an alarm of fire. Based on the appearance of the citizens, this picture was most likely prior to the organization of the Vigilant Hose Company of Emmitsburg in 1884. (Courtesy VHC.)

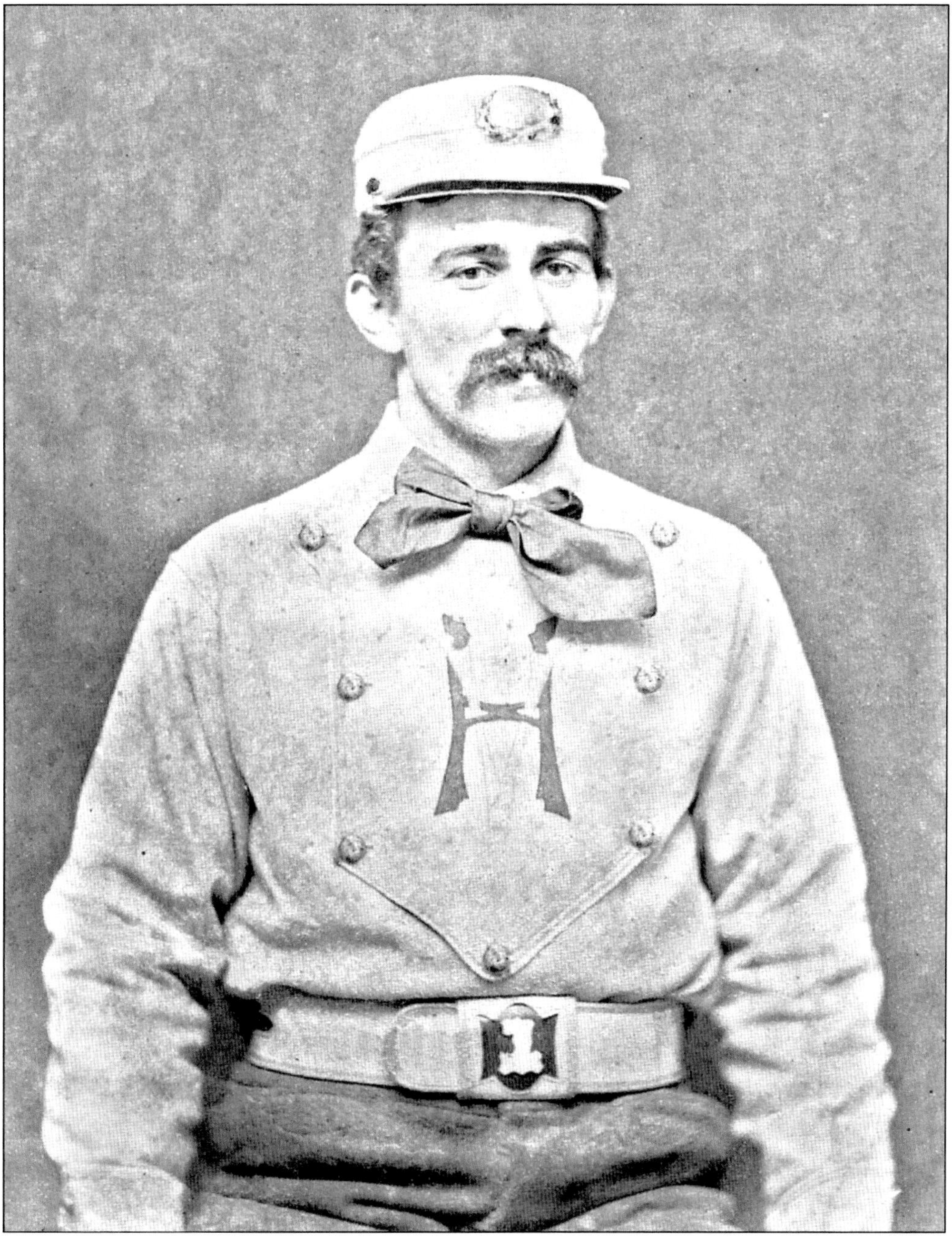

The first president of the Vigilant Hose Company of Emmitsburg was Mr. Edward Hoke Rowe. He is shown in the formal parade uniform of the Vigilant Hose Company. (Courtesy VHC.)

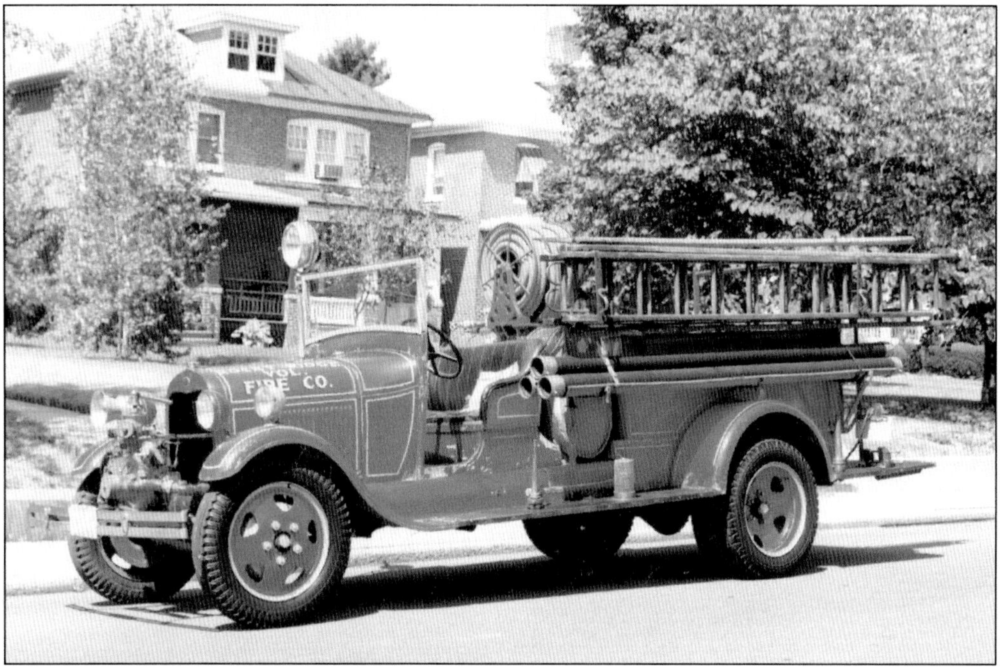

The 1929 Model A Ford pumper, now owned by the Rocky Ridge Volunteer Fire Company, was the first motorized engine purchased by the Vigilant Hose Company. This engine, purchased for $3,185, was proudly displayed on the cover of the 1936 Frederick County Firemen's Association Convention program book, with the Vigilant Hose Company along the body and "Emmitsburg, Md." written on the hood. (Courtesy Warren Jenkins.)

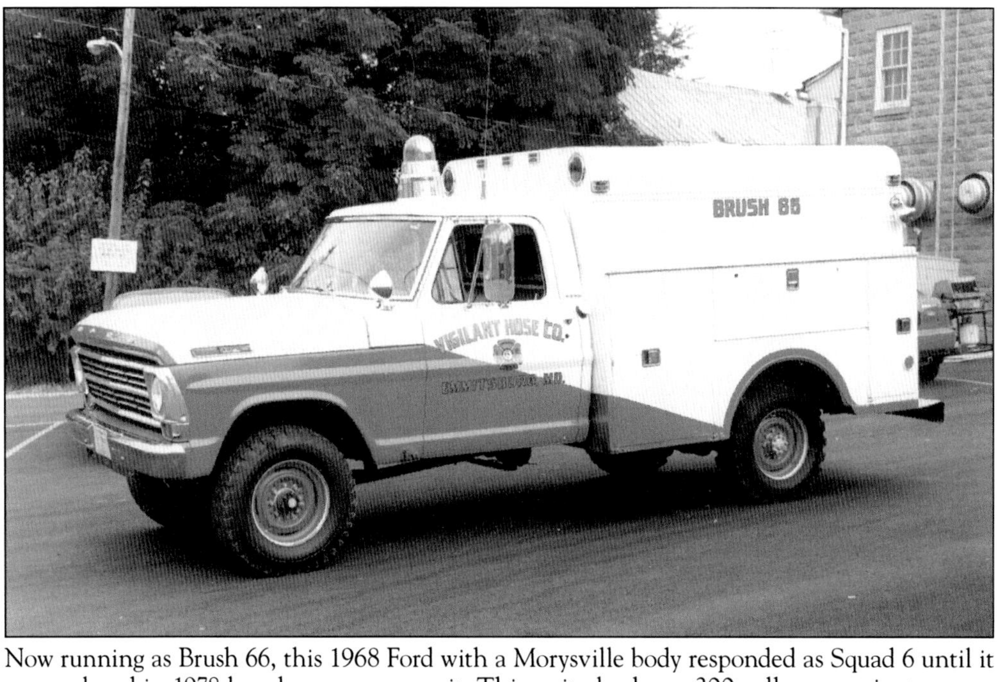

Now running as Brush 66, this 1968 Ford with a Morysville body responded as Squad 6 until it was replaced in 1978 by a heavy rescue unit. This unit also has a 300-gallon-per-minute pump and carries 300 gallons of water. (Courtesy Warren Jenkins.)

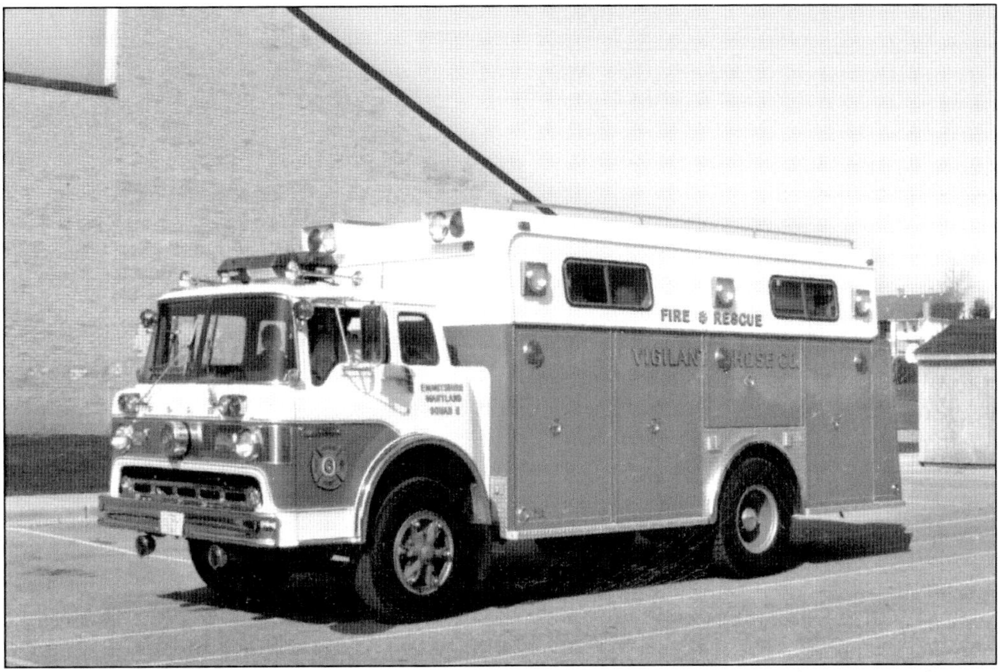

The first heavy rescue unit purchased by the Vigilant Hose Company was the 1978 Ford C900 Swab heavy duty rescue squad designated Squad 6. This unit responded to auto accidents and rescue throughout northern Frederick County and lower Adams County, Pennsylvania, as well as parts of Carroll County, Maryland, and Franklin County, Pennsylvania. (Courtesy Warren Jenkins.)

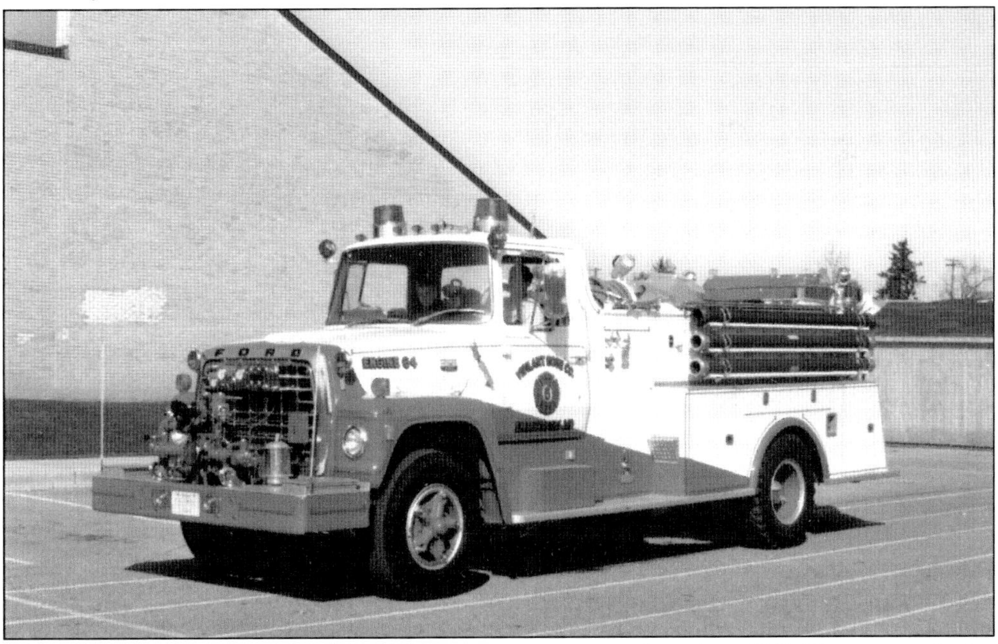

Engine 64 was a front-mounted Ford/American pumper purchased in 1971. This engine had a 750-gallon-per-minute pump and carried 1,000 gallons of water. The engine was donated to a rural area of Arkansas and is still in use in that community. (Courtesy Warren Jenkins.)

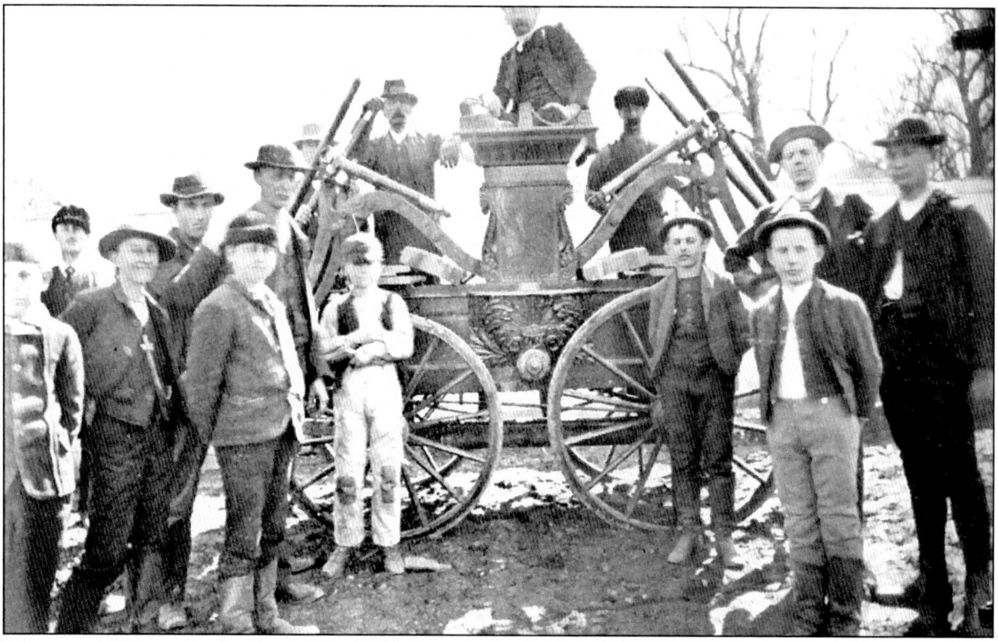

The year 1803 marks the first record of fire protection in Middletown. At one point, two companies, the Independents and Juniors, served the town, sometimes in open competition. Around 1850, the ladies of the community held a fair to raise money to purchase the first fire engine. The hand pumper remained in service to protect the community until at least 1894. (Courtesy Middletown Volunteer Fire Company [MVFC].)

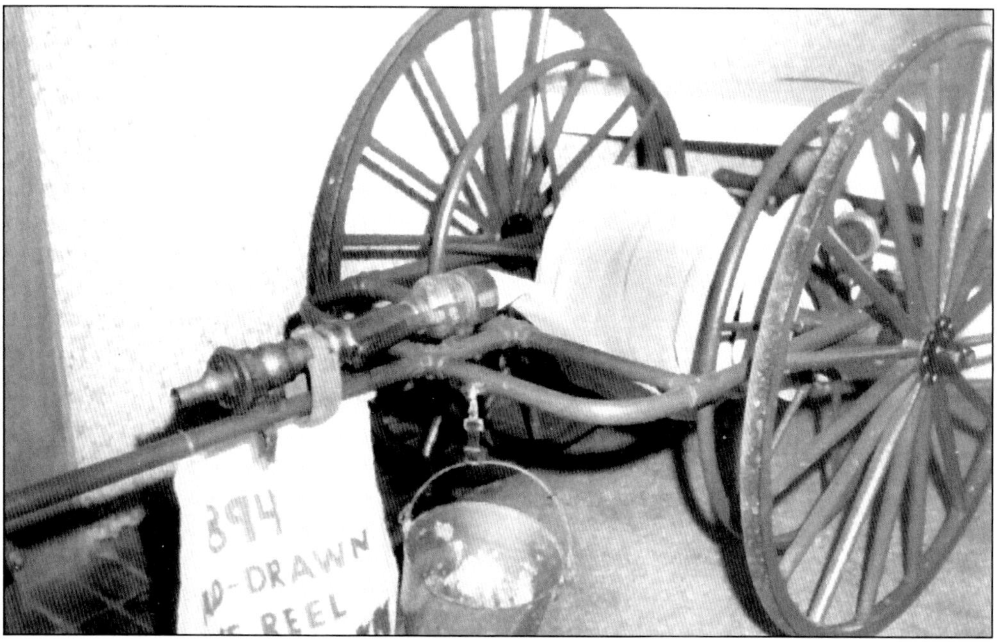

After a disastrous fire at the tannery in 1892, the town burgess and commissioners called a meeting to discuss town fire protection. In 1894, this hand-drawn hose reel was acquired to supplement the hand pumper, and a formal fire company was organized. Fire service has been continuous in Middletown since that time. (Courtesy MVFC.)

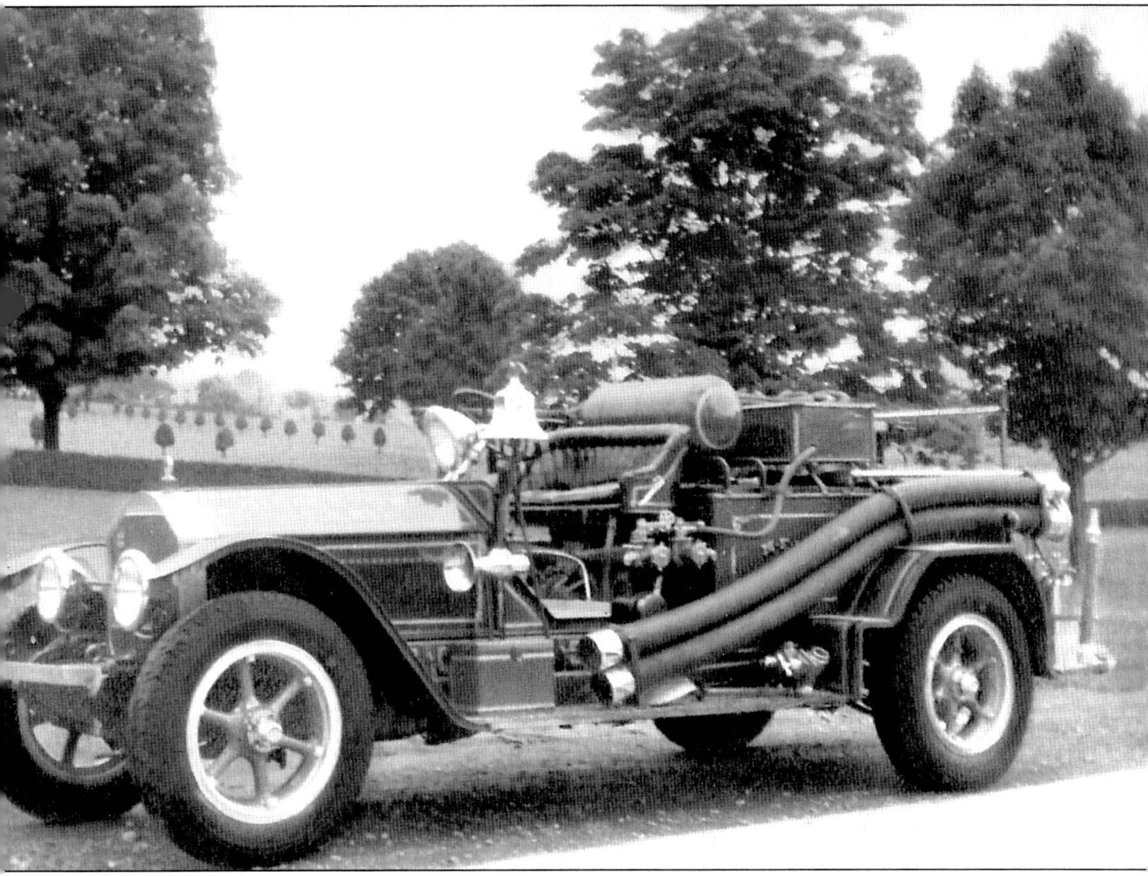

The first motorized apparatus was purchased in 1923 with this unique Kissel pumper. This chain-driven engine was housed in Lee Shafer's building across from the present fire station until room was made at Memorial Hall by removing walls on the east side of the building. (Courtesy MVFC.)

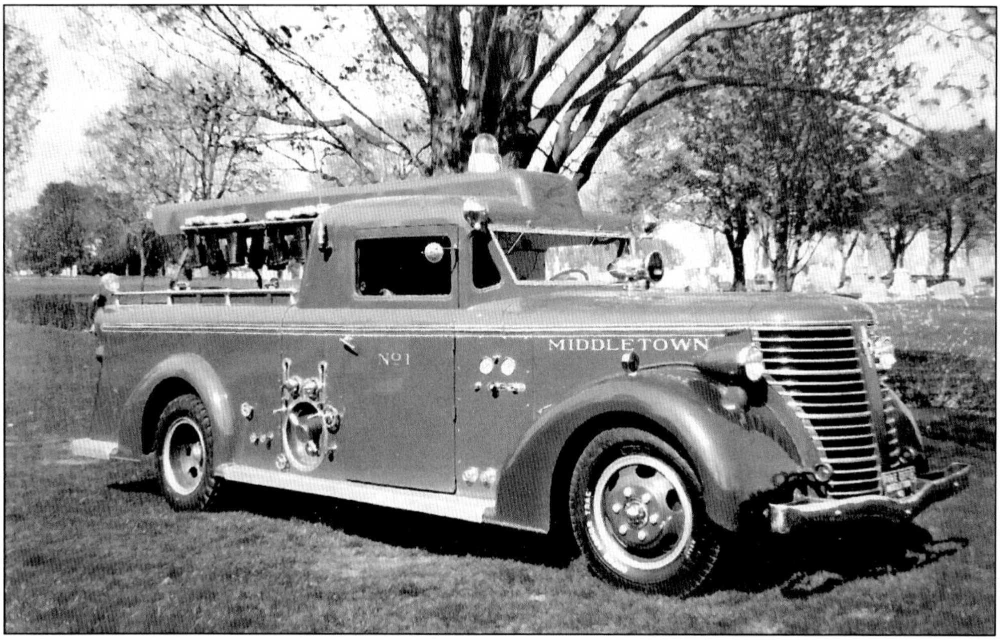

This sleek 1939 American LaFrance engine had a unique roof-mounted ladder. Originally called Engine No. 1, it was the first Engine 71 upon the issuance of company numbers in the county. It was sold to the Lewistown District Volunteer Fire Department for a cost of $450 to help that department organize. It responded to a working house fire in Mountaindale before officially being placed in service in Lewistown. (Courtesy MVFC.)

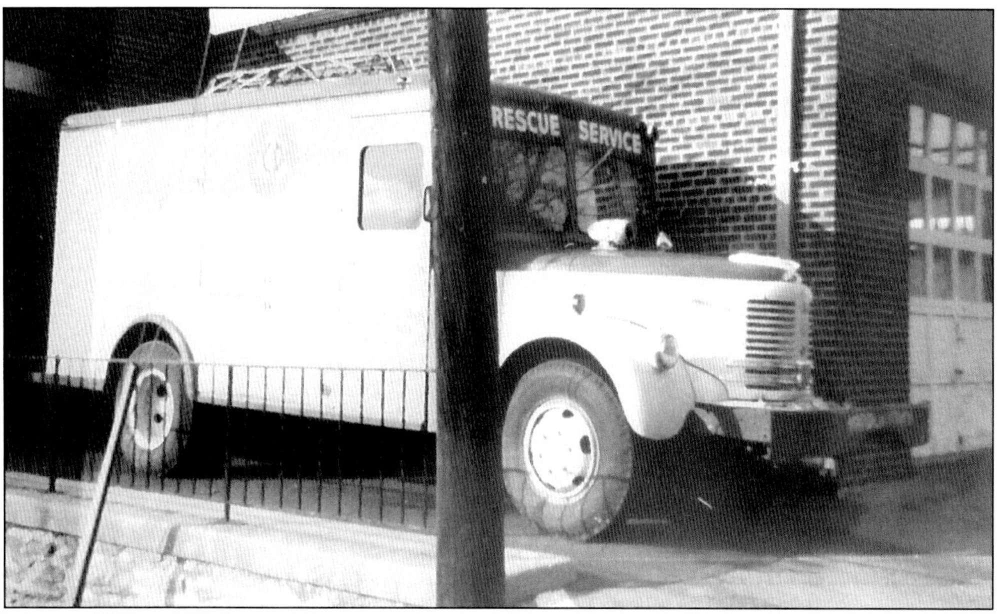

The Middletown Volunteer Fire Company received two units from Civil Defense. Shown is Squad 7, a Civil Defense rescue truck stationed at the fire company. The company also received a pumper from Civil Defense. Both units were owned by Civil Defense. The squad was replaced by Civil Defense in 1976 with another squad that was ultimately purchased by the fire company for $1. (Courtesy MVFC.)

Though the model is unknown, Middletown was the first fire company in the county to purchase a chief's car. This car, believed to be from around 1935, was equipped with a siren on the fender. (Courtesy MVFC.)

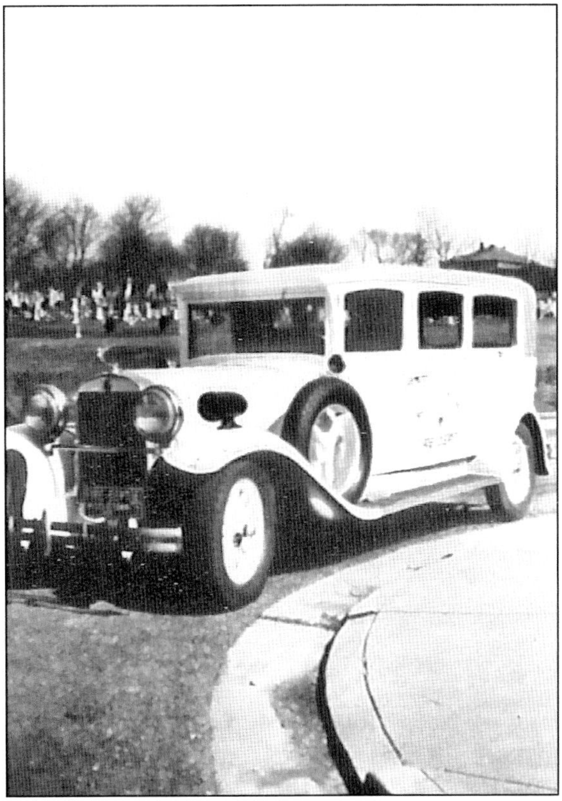

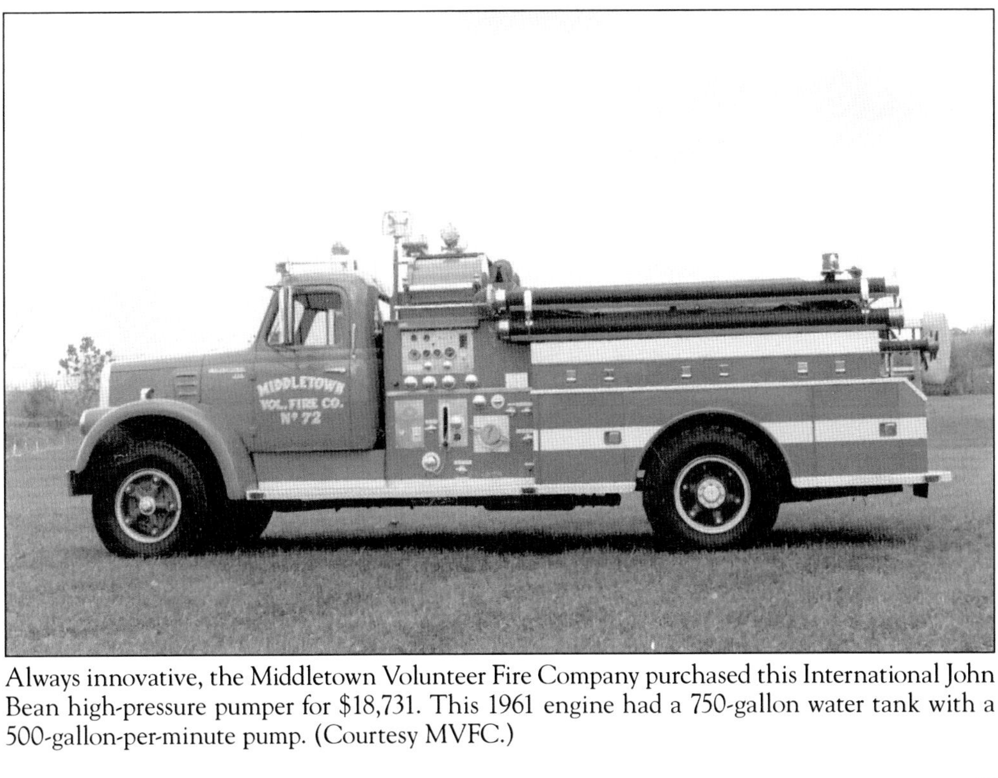

Always innovative, the Middletown Volunteer Fire Company purchased this International John Bean high-pressure pumper for $18,731. This 1961 engine had a 750-gallon water tank with a 500-gallon-per-minute pump. (Courtesy MVFC.)

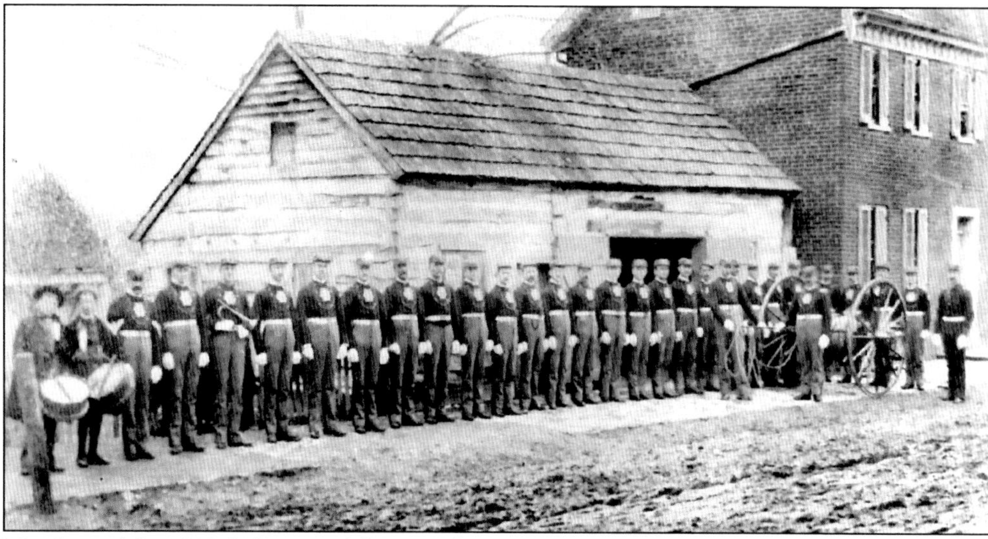

Members of the 1888 drill team of the Guardian Hose Company stand in front of the first engine house with the hose reel. Shown from left to right are Victory Cramer, Harry Cover, Charles Black, William Sefton, Martin Foreman, Edwin Gernard, John Cover, William Black, Alfred Weller, Charles R. Eyler Sr., Samuel B. Bennett, J. Irwin Mackley, Martin Buzzard, James Stocks, Edward L. Root, Edward Zeck, Harry Gaugh, M. Luther Ceager, McClellan Baltzell, J. Wesley Creeger, Morris L. Rouzer, Henry W. Henshaw, John A. Stull, Charles H. Osler, George Wireman, Charles M. Mackley, Elmer E. Black, Charles M. Root, Calvin C. Fox, Robert E. Creeger, Henry Bennett, and Frank Stokes. (Courtesy Thurmont Historical Society.)

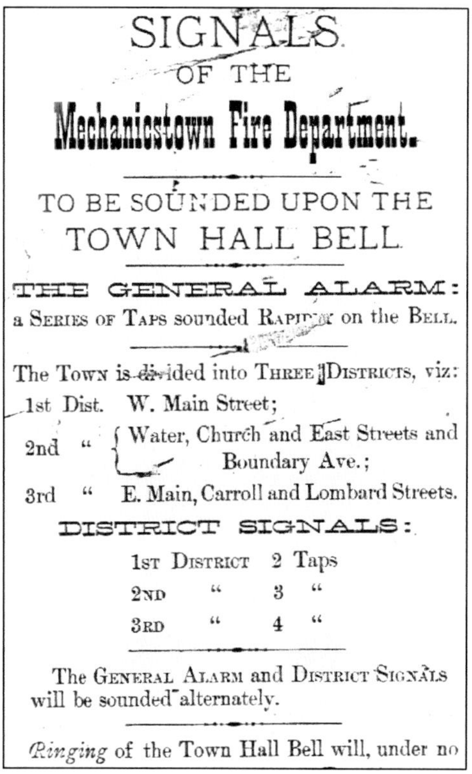

SIGNALS
OF THE
Mechanicstown Fire Department.

TO BE SOUNDED UPON THE
TOWN HALL BELL.

THE GENERAL ALARM:
a Series of Taps sounded Rapidly on the Bell.

The Town is divided into Three Districts, viz:

1st Dist. W. Main Street;
2nd " Water, Church and East Streets and Boundary Ave.;
3rd " E. Main, Carroll and Lombard Streets.

DISTRICT SIGNALS:

1st District	2 Taps
2nd "	3 "
3rd "	4 "

The General Alarm and District Signals will be sounded alternately.

Ringing of the Town Hall Bell will, under no

The signals of the Mechanicstown Fire Department are displayed on this poster by the members of the Guardian Hose Company. The alarm was sounded using the bell on the town hall located atop the structure also used as the fire company. The town name changed to Thurmont in 1893. (Courtesy Thurmont Historical Society.)

The town hall and fire station located on Church Street were built in 1890. This structure was used by the Guardian Hose Company until it was razed in 1950 to make way for a new fire station. At the printing of this book, the Guardian Hose Company is planning to build a new fire station on this site. (Courtesy Thurmont Historical Society.)

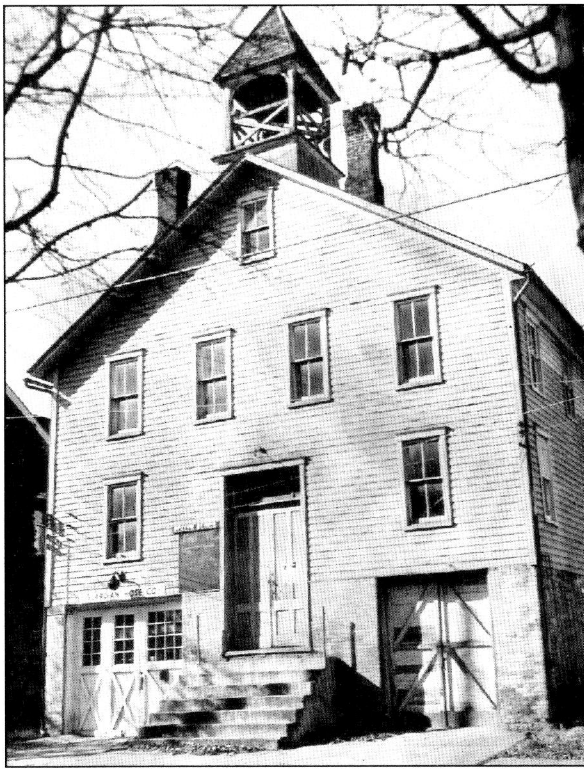

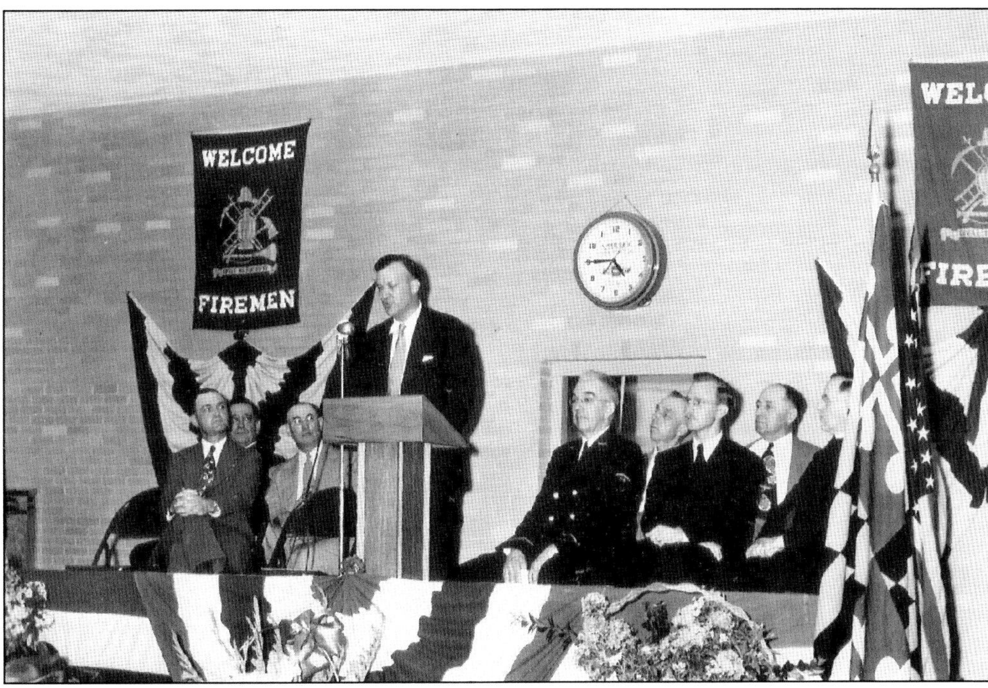

Upon the completion of the new fire hall, a dedication was held to celebrate the occasion. Here Gov. Theodore McKeldin addresses the dignitaries and members present at the ceremony. (Thurmont Historical Society.)

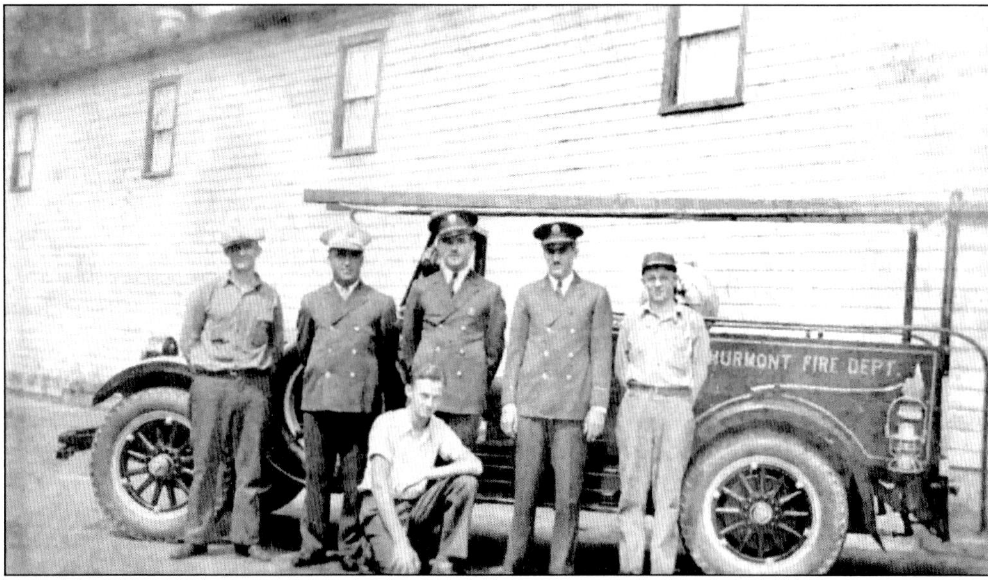

The fire company purchased a 1925 Hudson Sedan automobile and converted it into a fire engine. Shown with the engine after the conversion are, from left to right, as follows: (kneeling) Mr. Kerschner; (standing) Libert Weddle, Ernest Hammaker, Harry O. Miller, John Creager, and Roy Wisotzkey. (Courtesy Thurmont Historical Society.)

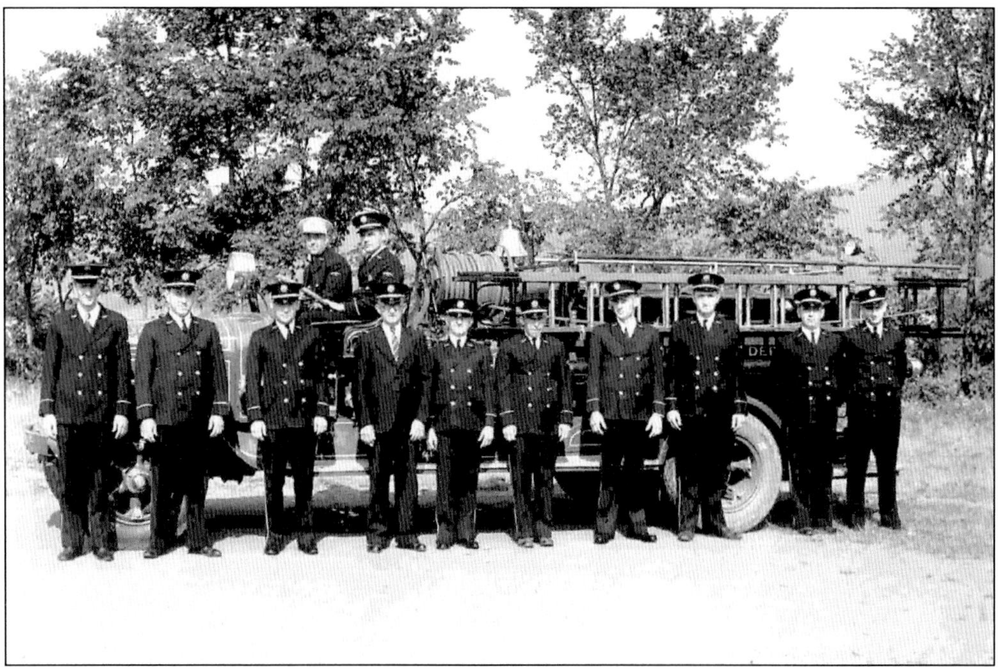

Members of the Guardian Hose Company pose in full dress uniform in front of the new 1927 chain-driven Mack fire engine. This engine is now on display at the Fire Museum of Maryland, where it is on loan from the fire company. Shown from left to right are (seated) Chief John Creager and driver Harry O. Miller; (standing) Earl Creager, Lee Martin, Charles R. Ambrose, Samuel B. Bennett, Erasmus Eyler, Norman Shaffer, Taymond Knott, Lloyd Mackley, George Black, and James Eyler. (Courtesy Thurmont Historical Society.)

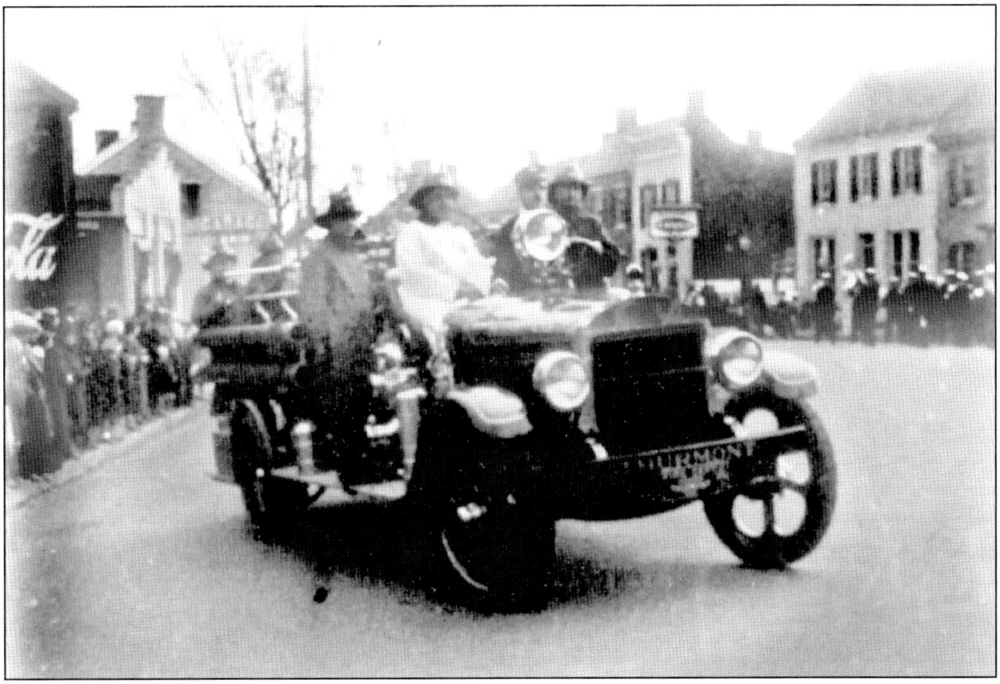

Members of the Guardian Hose Company in turnout gear approach Seventh Street in a Frederick parade. The officer is distinguished by the white gear, as the driver wears his dress uniform. (Courtesy Thurmont Historical Society.)

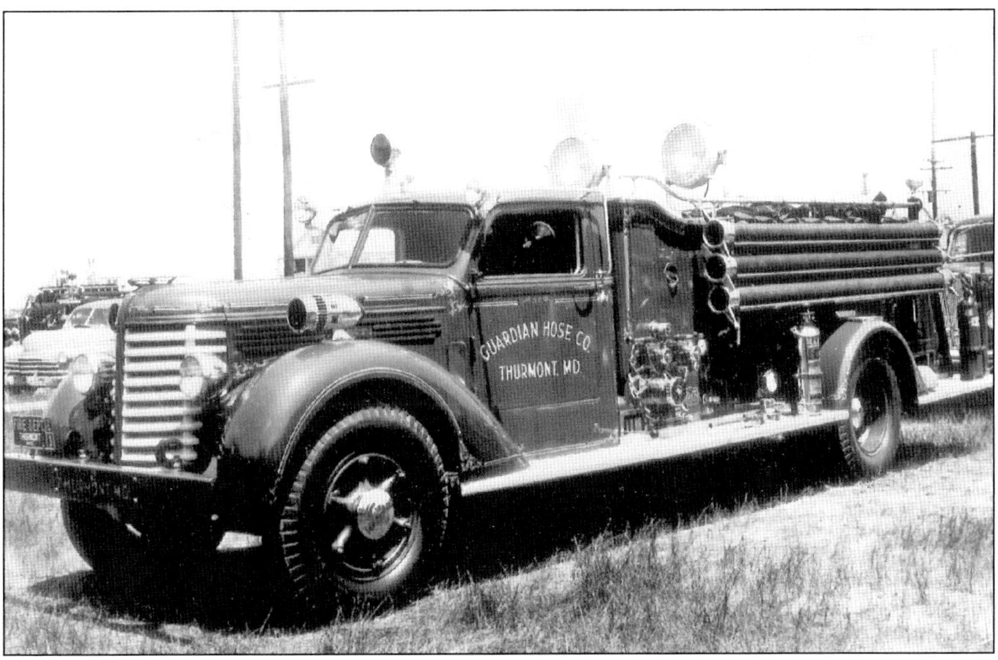

The Guardian Hose Company continued to meet the needs of the community with the purchase of this 1945 Diamond-T engine. This engine remained in service for over 30 years. Note the large floodlights mounted on the engine. (Courtesy Thurmont Historical Society.)

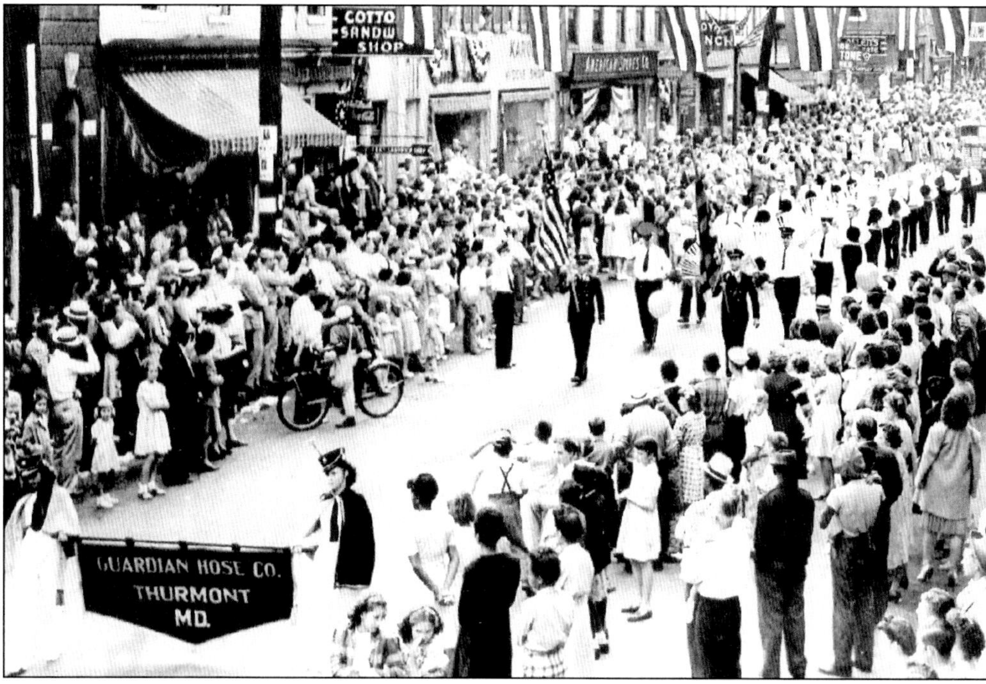

The marching unit of the Guardian Hose Company is followed by the Diamond-T engine in the 1947 Maryland State Firemen's Association Parade in Frederick. This was the first parade held by the MSFA after World War II. (Courtesy Thurmont Historical Society.)

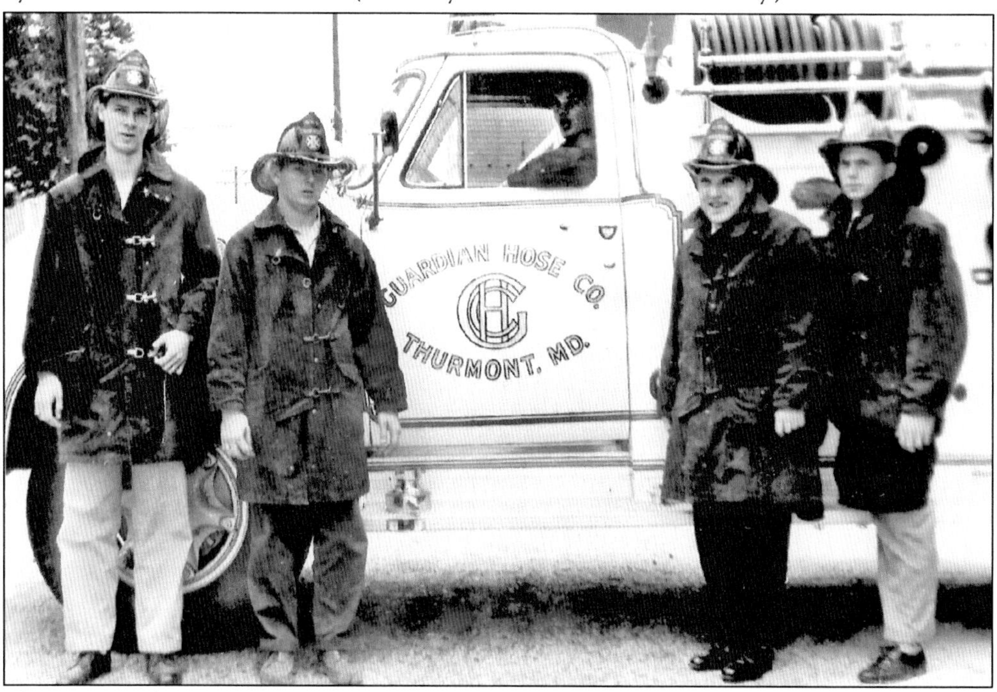

The 1954 Ford pumper was nearly new when the members posed in front of the fire station. Shown from left to right are Carroll Brown, Dick Wilhide, driver Bob Wilhide, Bill Stitely, and Pete Welch. (Courtesy Thurmont Historical Society.)

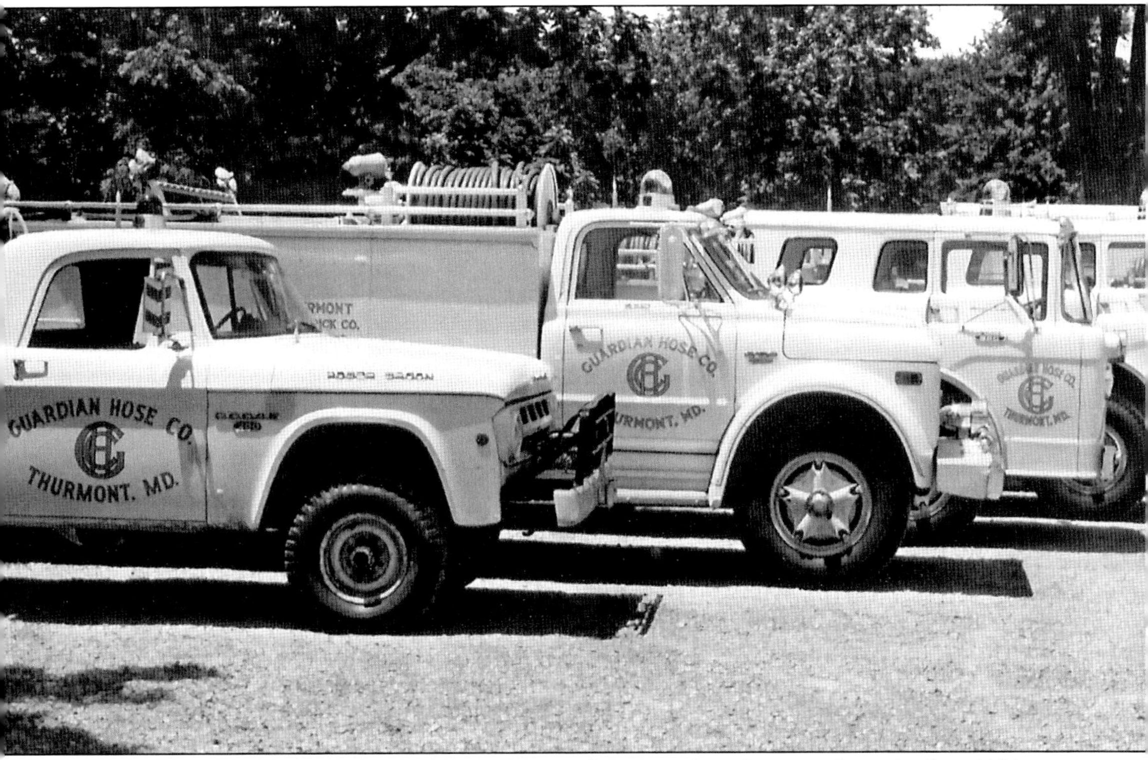

The equipment of the Guardian Hose Company is shown in this photograph in the late 1970s. From left to right, Brush 105 is a 1968 Dodge Power Wagon Brush Truck, Tanker 10 is a 1970 Chevrolet/Oren Tanker, Engine 102 a 1954 Ford/Oren, Engine 101 is a 1967 Ford/Hahn Pumper, Engine 103 is a 1977 Ford/Hahn, and Squad 10 is a 1972 Ford/Swab Heavy Rescue Squad. (Courtesy Thurmont Historical Society.)

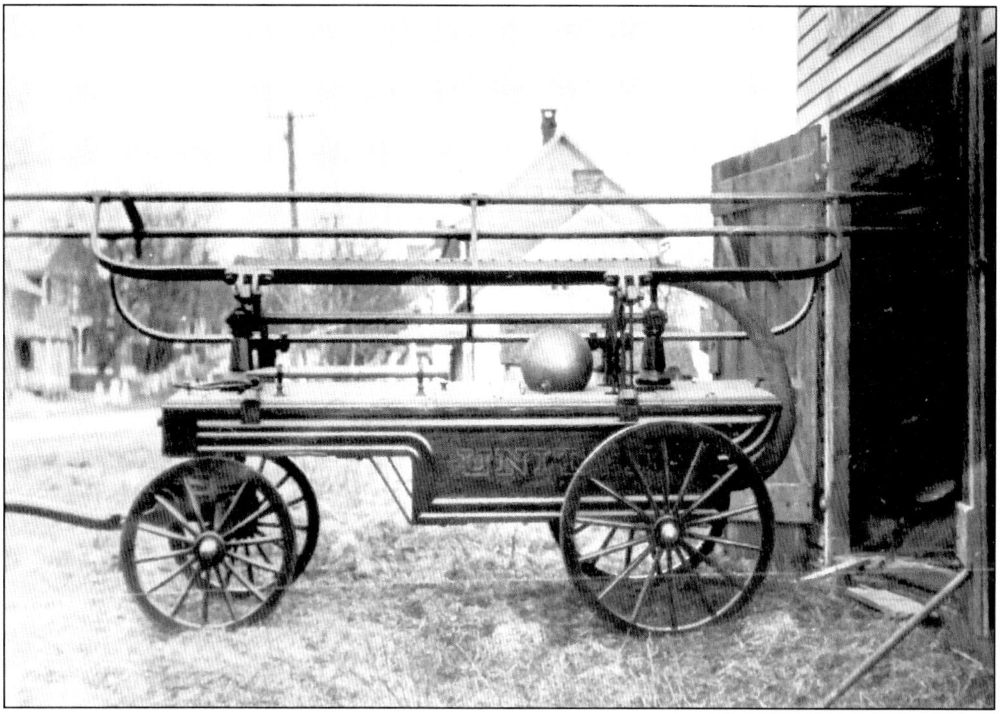

The first fire engine in Libertytown was the "Old Lady," a hand-tub pumper purchased from the United Steam Fire Engine Company No. 3 of Frederick. The engine is shown here at the original fire station located in the center of Libertytown. This hand pumper was recently purchased by the Granite Hand Tub Society in New Hampshire and will be completely restored and once again operable. (Courtesy United Fire Company.)

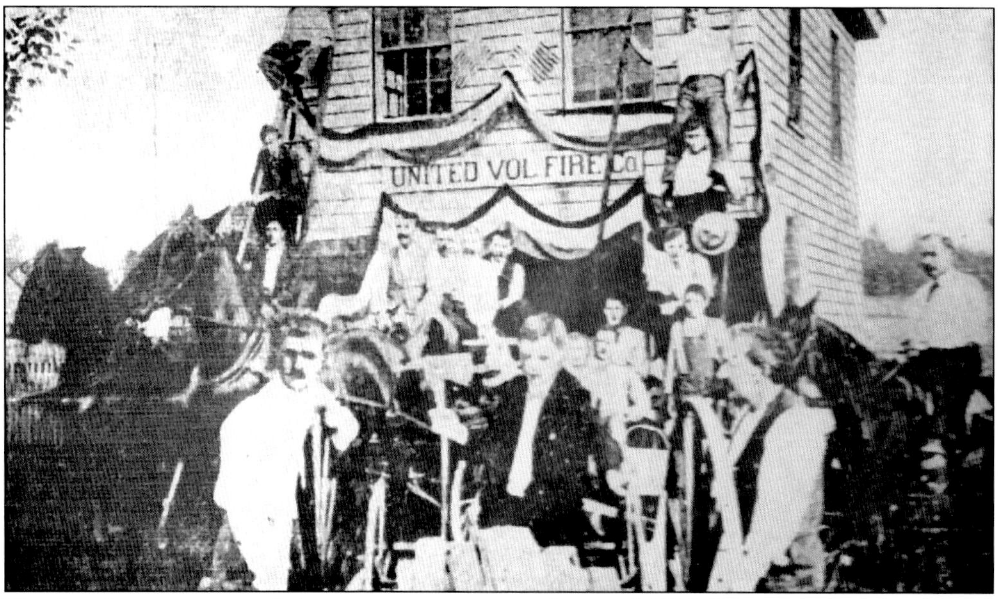

The first fire company in Libertytown was organized around 1890. The company assumed the name of the previous owner displayed on the side of the engine, a squirrel-tail pumper with a hand-tub trailer and hard sleeve that extended to the tub.

Two

FIFTY YEARS OF PROGRESS
1900 to 1950

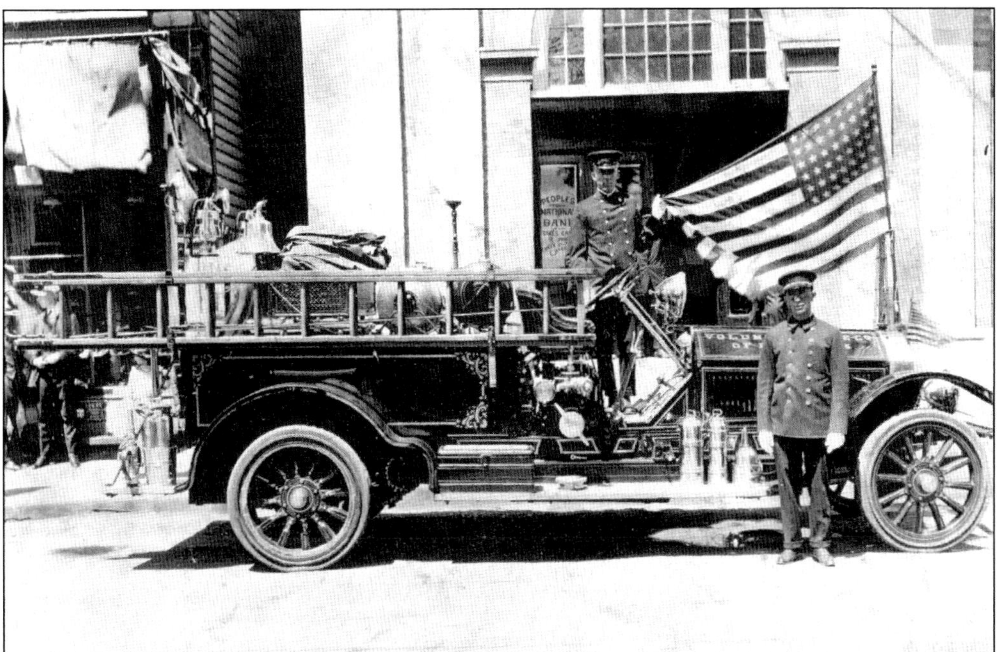

In 1922, the Brunswick Volunteer Fire Company purchased an American LaFrance engine. The engine is shown in front of the Peoples National Bank on Potomac Street in Brunswick. This engine was sold to Mr. Leonard King, past president of the Maryland State Firemen's Association, and was recently bought back by the fire company and is being restored. (Courtesy Brunswick Volunteer Fire Company [BVFC].)

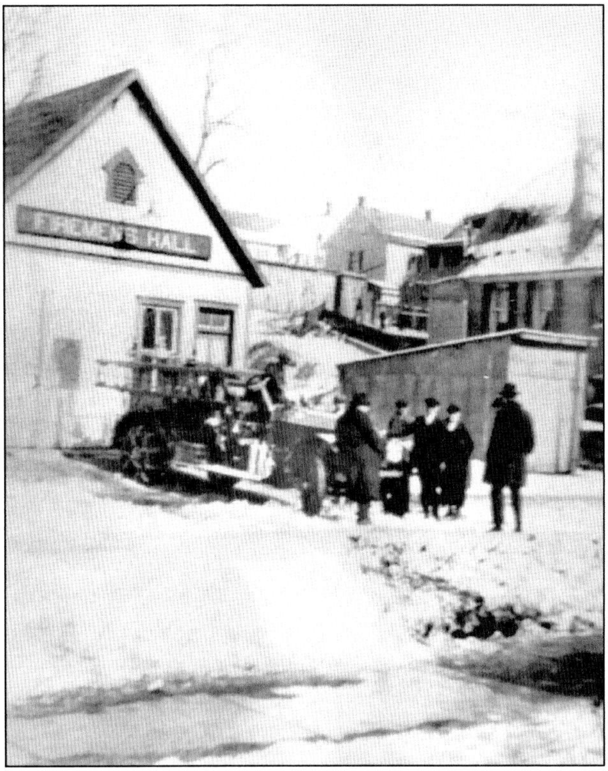

Though the first official fire protection in Brunswick was organized in 1897, the Brunswick Volunteer Fire Company did not come into existence until May 10, 1910. The original fire house, Firemen's Hall, was built on North Delaware Avenue and was turned over to the company by the City of Brunswick in 1928. (Courtesy BVFC.)

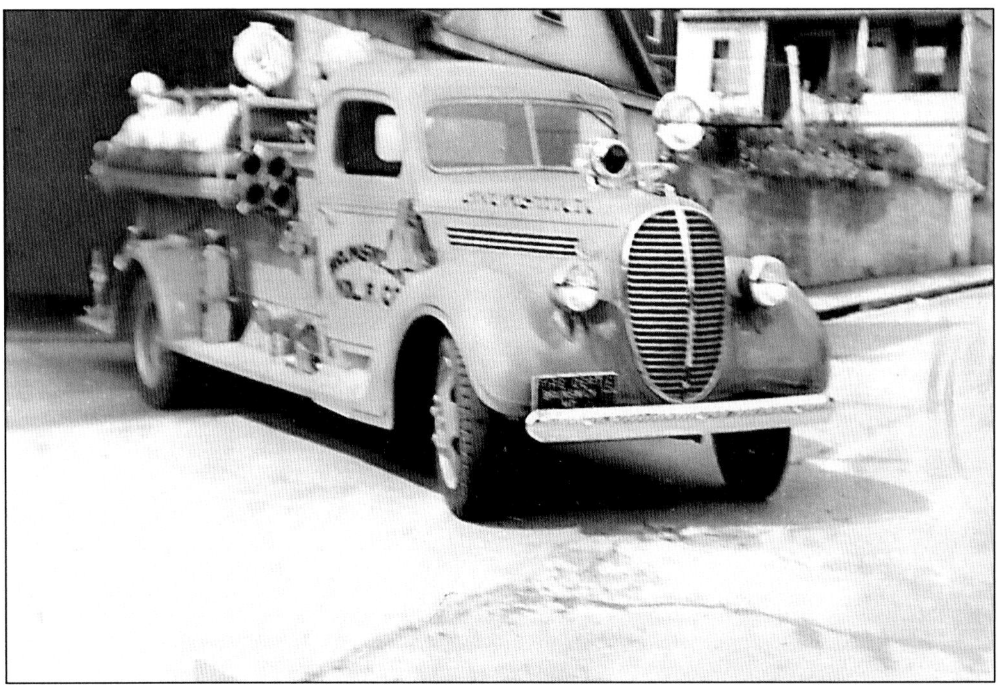

In 1938, a 1938 Ford Howe fire engine, equipped with a 350-gallon-per-minute piston pump, was purchased to respond to calls outside the limits of Brunswick. A most interesting note is that the engine was robin's egg blue. (Courtesy Tom Watson.)

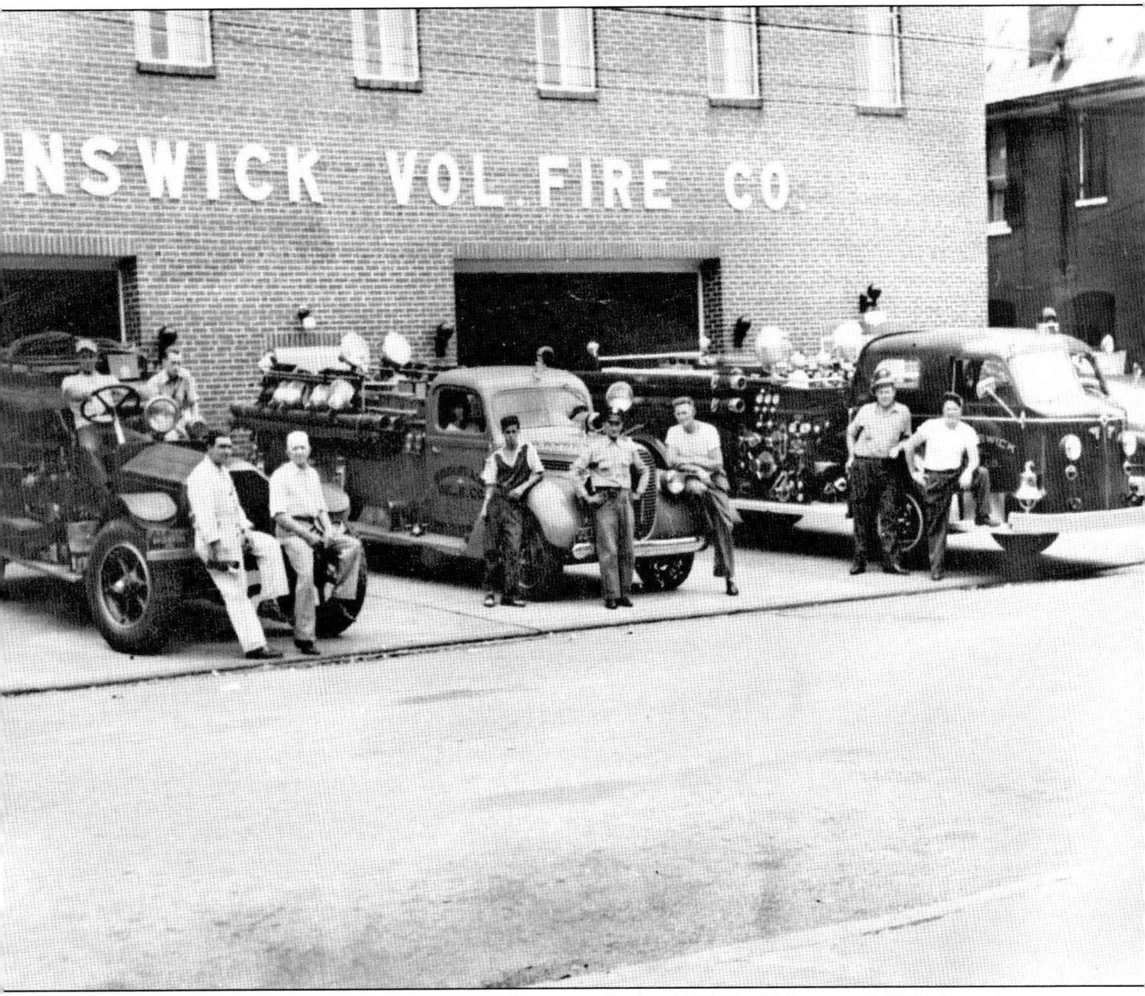

The new 1953 American LaFrance engine is shown with the 1938 Ford Howe and 1922 American LaFrance in front of the new Brunswick Fire Hall. This building was designed with a huge auditorium-type hall on the second floor to accommodate some of the country's most highly regarded entertainers. (Courtesy BVFC.)

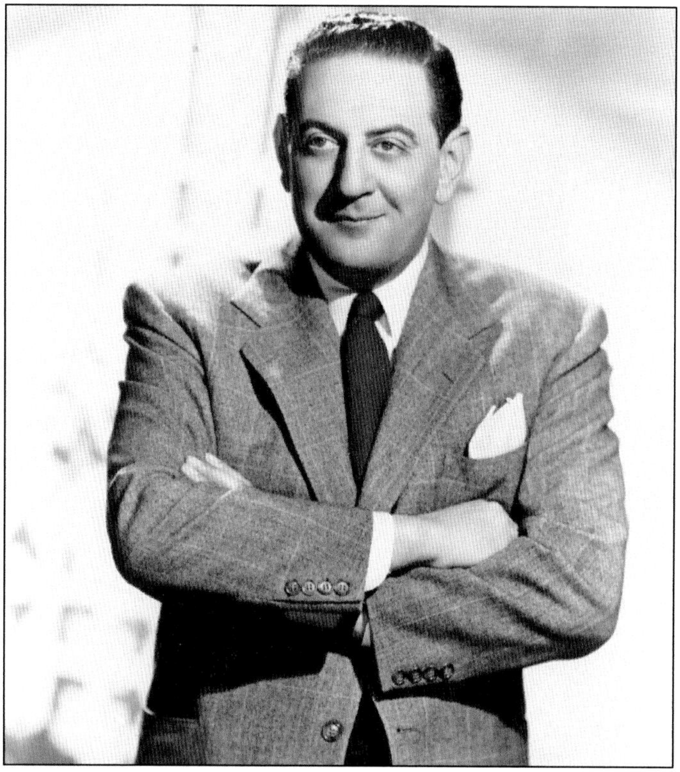

Guy Lombardo and His Royal Canadians were among the national entertainers to grace the Brunswick Fire Hall. The facility was noted throughout the Washington, D.C., area for its high-quality entertainment. Fire Chief Harvey "Sonny" Cannon owned a sound company and was able to attract some of the big names of the 1950s and 1960s to Brunswick. (Courtesy BVFC.)

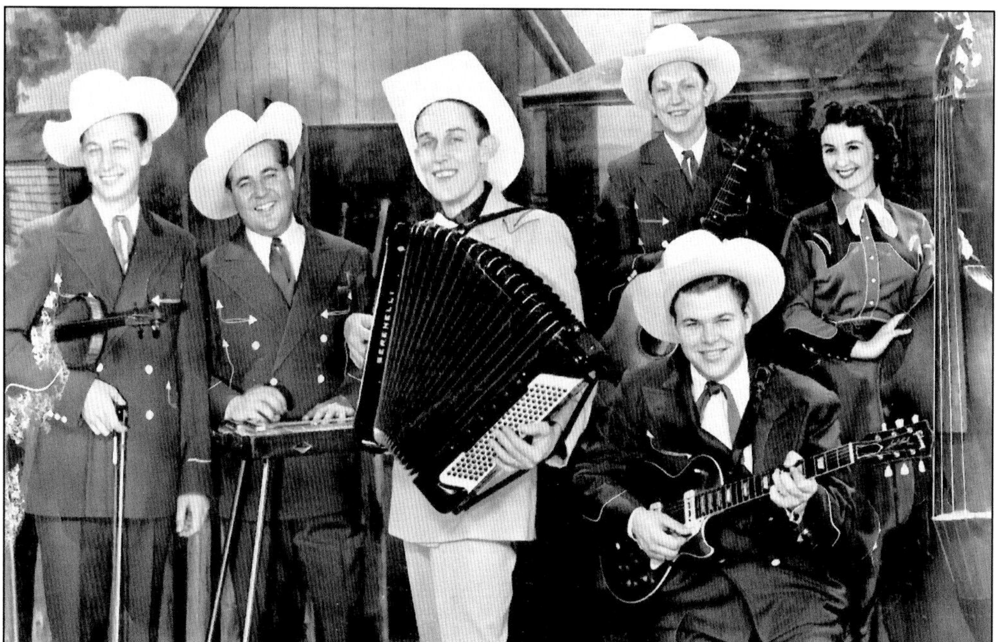

Country legends Jimmy Dean, Patsy Cline, and Roy Clark can all thank the Brunswick Fire Hall for helping launch their country careers. This photograph, given to Tom Watson by Jimmy Dean, shows, from left to right, Buck Ryan, Maurice Carroll, Jimmy Dean, Herbie Jones, Roy Clark, and Dale Turner. (Courtesy Tom Watson.)

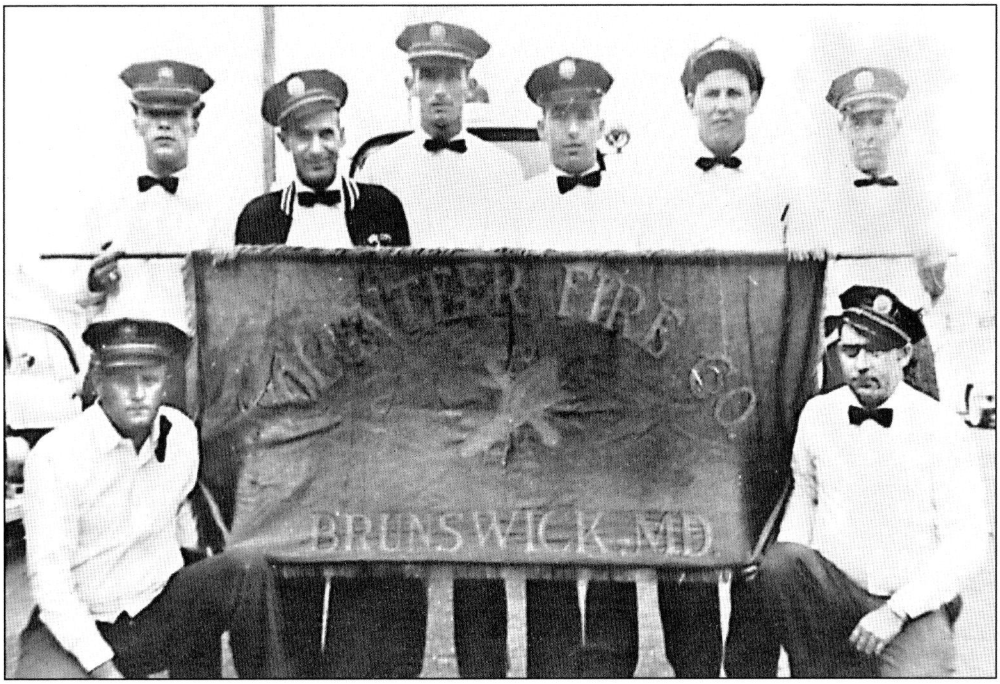

In this photograph from the 1950s, members of the Brunswick Volunteer Fire Company are, from left to right, (seated) Freddie Moore and Bud Harrison; (standing) unidentified, Frank Miglio, Doug Frye, Clair Ebersole, Webb Bennett, and Pete Virts. (Courtesy BVFC.)

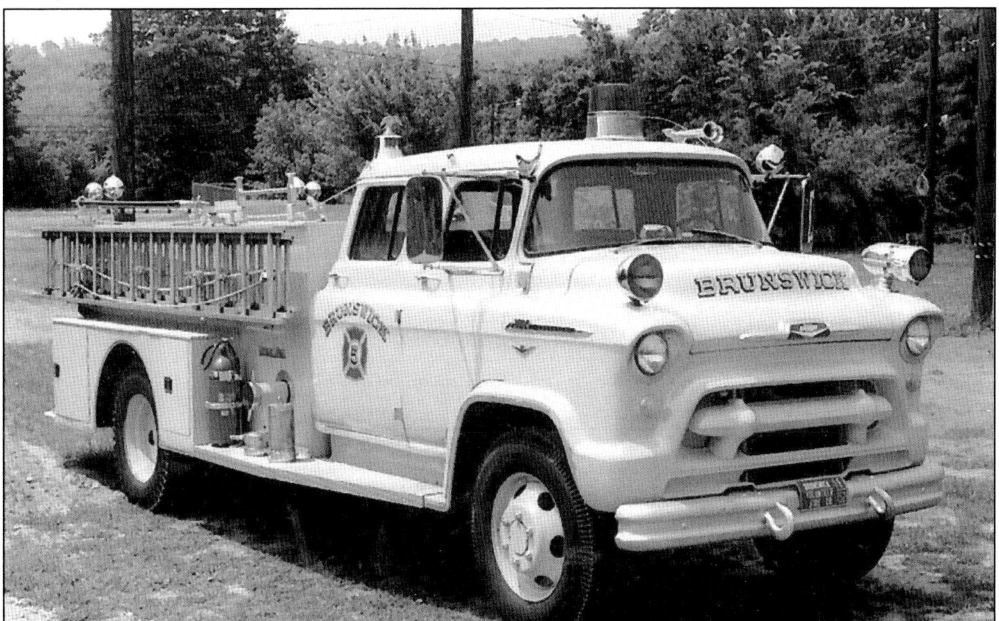

In 1957, the Brunswick Volunteer Fire Company purchased this Chevrolet pumper with a four-door cab to safely transport firefighters to emergencies. This engine was known as "Jimmy Dean's Engine," since money raised at the Saturday night dances with Jimmy Dean and the Texas Wildcats helped to buy the truck. (Courtesy Warren Jenkins.)

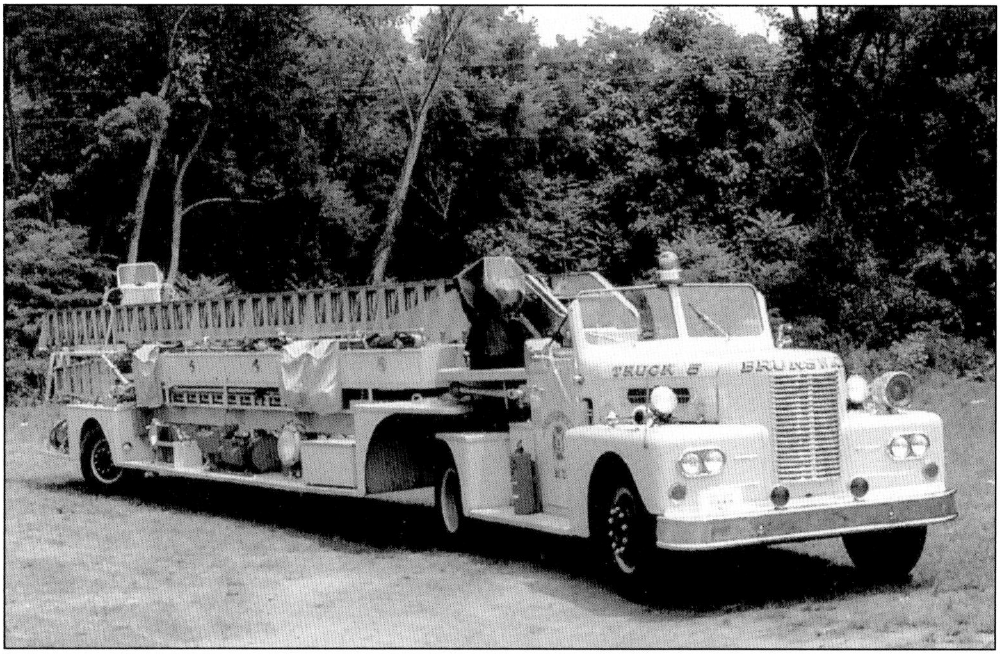

In 1982, a 1960 Pirsch 100-foot aerial ladder truck was purchased from the D.C. Fire Department for $100. This ladder truck has a tractor-drawn tiller truck requiring a driver on the front and back. A building was built on the carnival grounds near the Acme Market to house the unit. (Courtesy Warren Jenkins.)

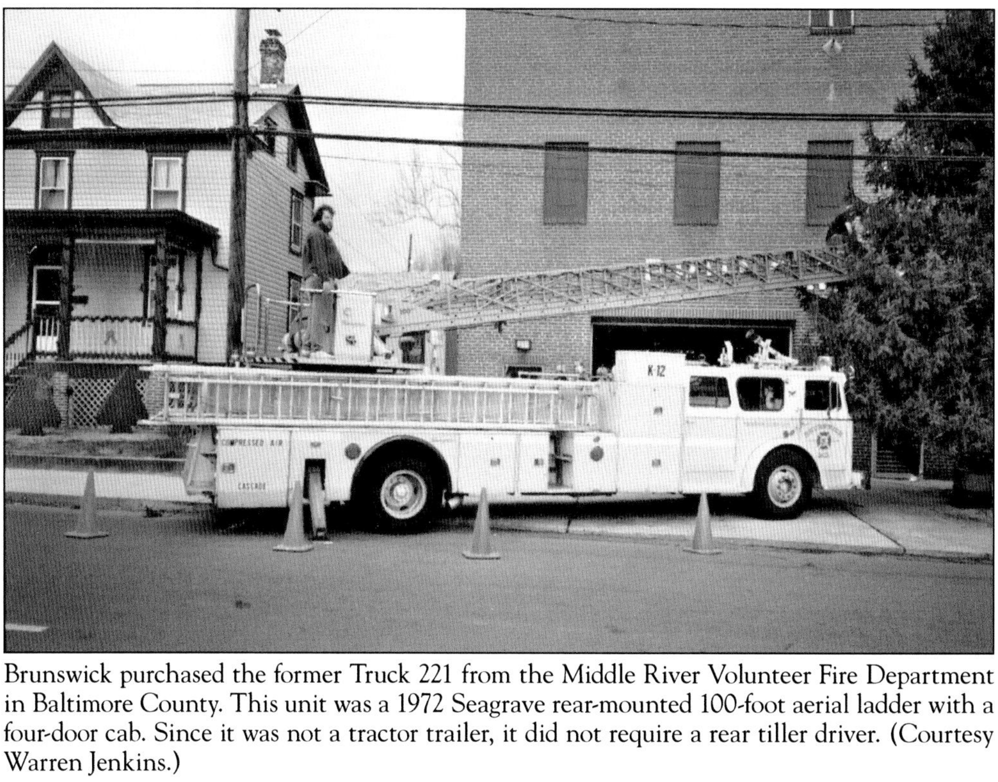

Brunswick purchased the former Truck 221 from the Middle River Volunteer Fire Department in Baltimore County. This unit was a 1972 Seagrave rear-mounted 100-foot aerial ladder with a four-door cab. Since it was not a tractor trailer, it did not require a rear tiller driver. (Courtesy Warren Jenkins.)

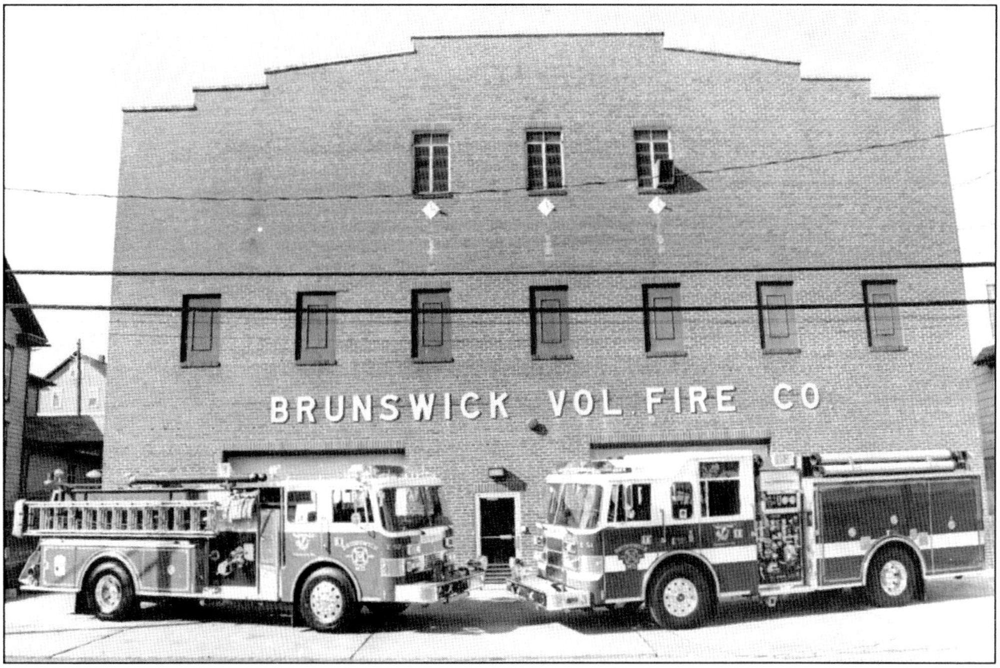

The latest equipment is shown in front of the Brunswick Volunteer Fire Company station on West Potomac Street in Brunswick. The fire company recently entered a cooperative effort with the Board of County Commissioners of Frederick County to build a new fire station off Route 17. Engine 52 on the left is a 1990 Pierce Lance; on the right, Rescue Engine 51 also a Pierce. (Courtesy BVFC.)

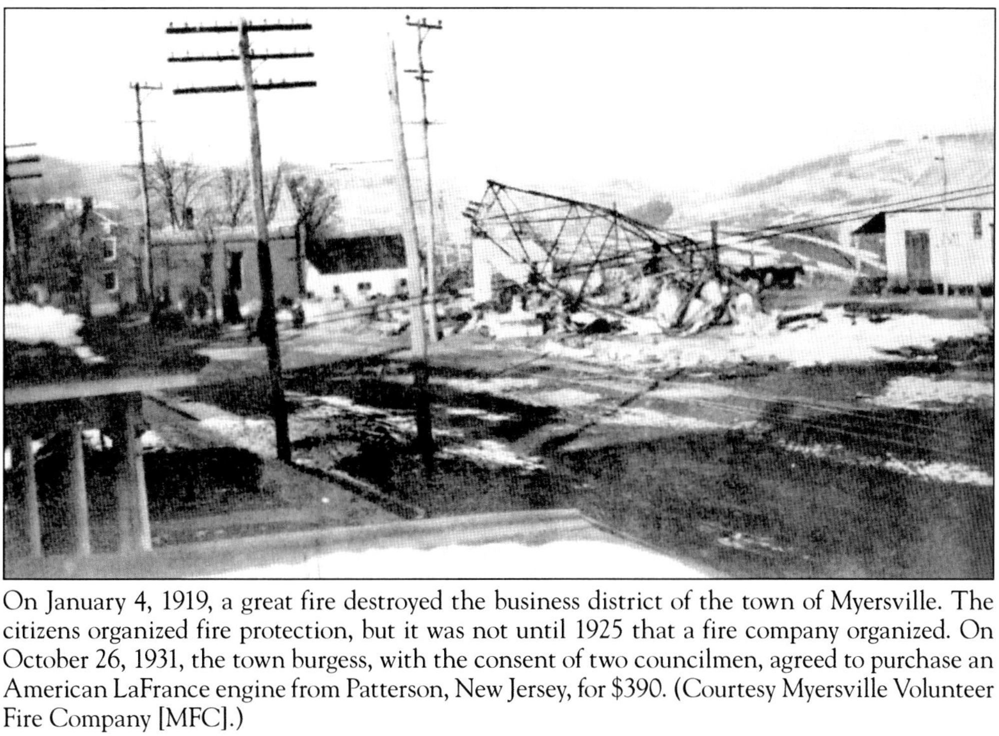

On January 4, 1919, a great fire destroyed the business district of the town of Myersville. The citizens organized fire protection, but it was not until 1925 that a fire company organized. On October 26, 1931, the town burgess, with the consent of two councilmen, agreed to purchase an American LaFrance engine from Patterson, New Jersey, for $390. (Courtesy Myersville Volunteer Fire Company [MFC].)

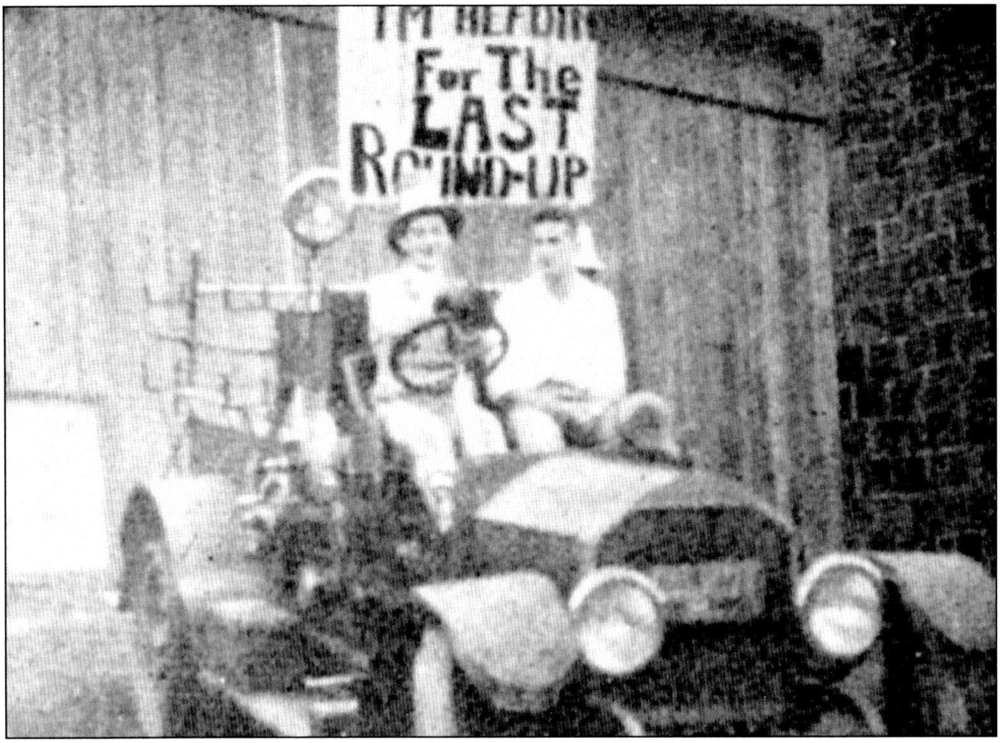

This 1917 American LaFrance was the first engine purchased by the Myersville Volunteer Fire Company to provide fire protection to the western environs of Frederick County. (Courtesy MFC.)

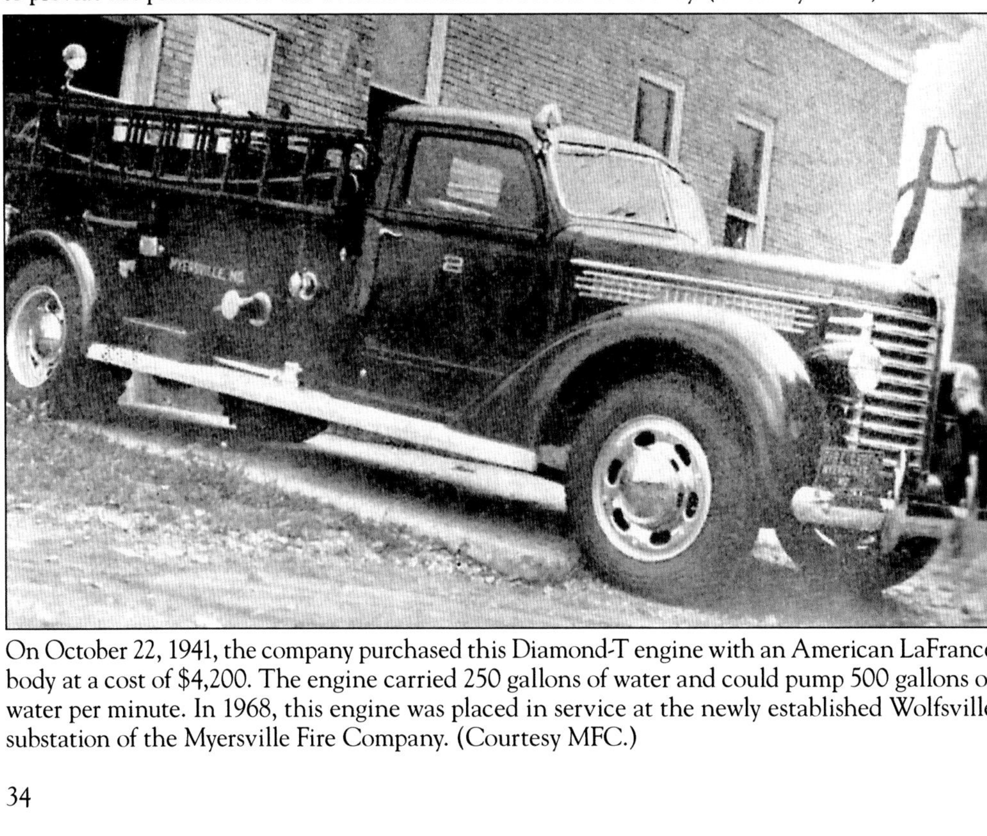

On October 22, 1941, the company purchased this Diamond-T engine with an American LaFrance body at a cost of $4,200. The engine carried 250 gallons of water and could pump 500 gallons of water per minute. In 1968, this engine was placed in service at the newly established Wolfsville substation of the Myersville Fire Company. (Courtesy MFC.)

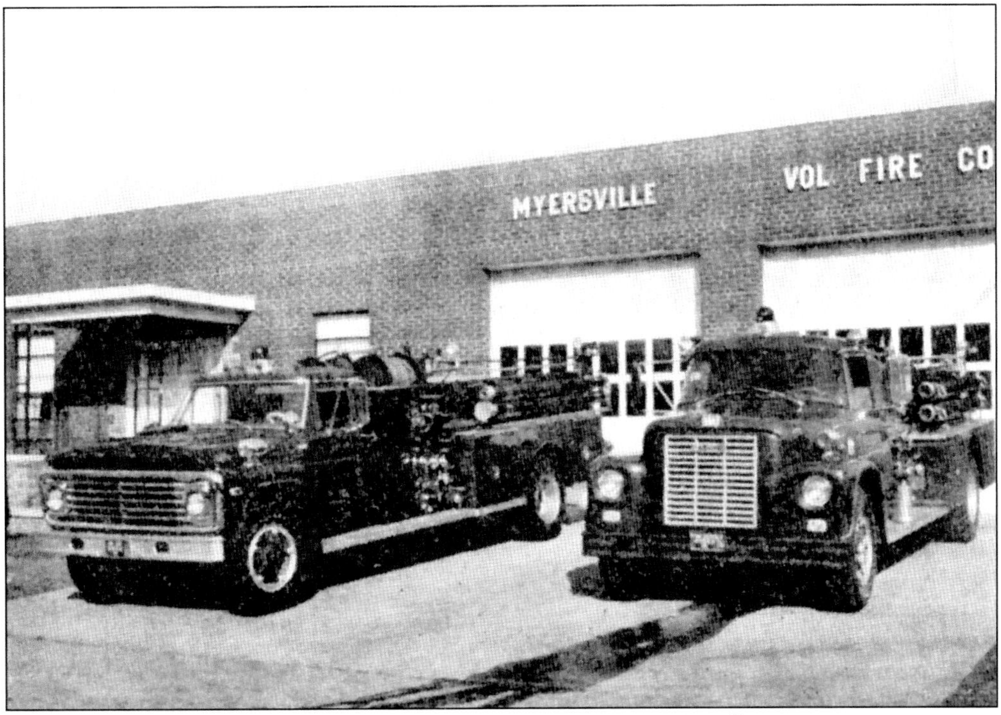

Shown in front of the fire station in Myersville are the 1962 Howe 750-gallon-per-minute pumper on an International chassis (right) and the 1968 Maxim engine on a Ford chassis. (Courtesy MFC.)

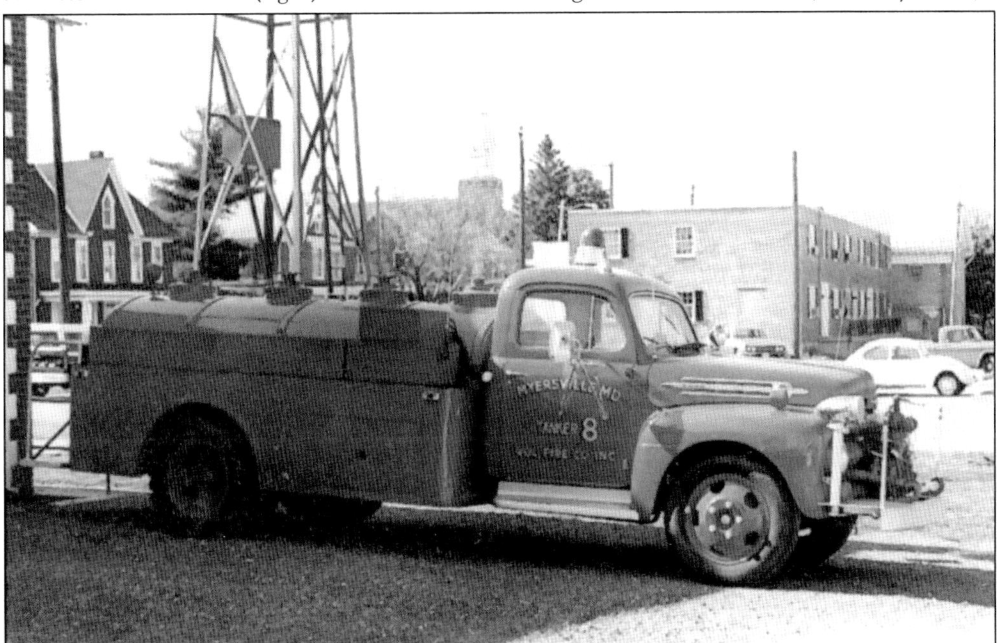

The fire company also placed in service this 1952 Ford tanker with a front-mounted pump. This unit responded as Tanker 8 and supplied additional water for fires in the rural areas. This tanker was originally owned by the Guardian Hose Company of Thurmont and was sold to Myersville for $1,000. (Courtesy MFC.)

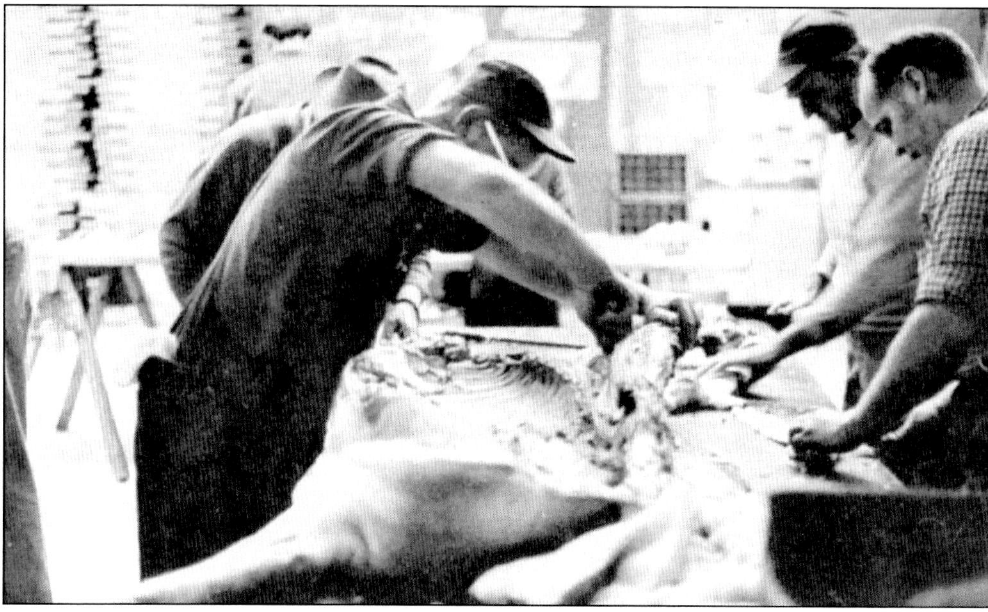

Fund-raising is a challenge for the volunteer fire company. Many local farmers would gather to hold a butchering of hogs and sell the meat to raise money. This 1970 picture shows members and local farmers cutting the meat for sale. (Courtesy MFC.)

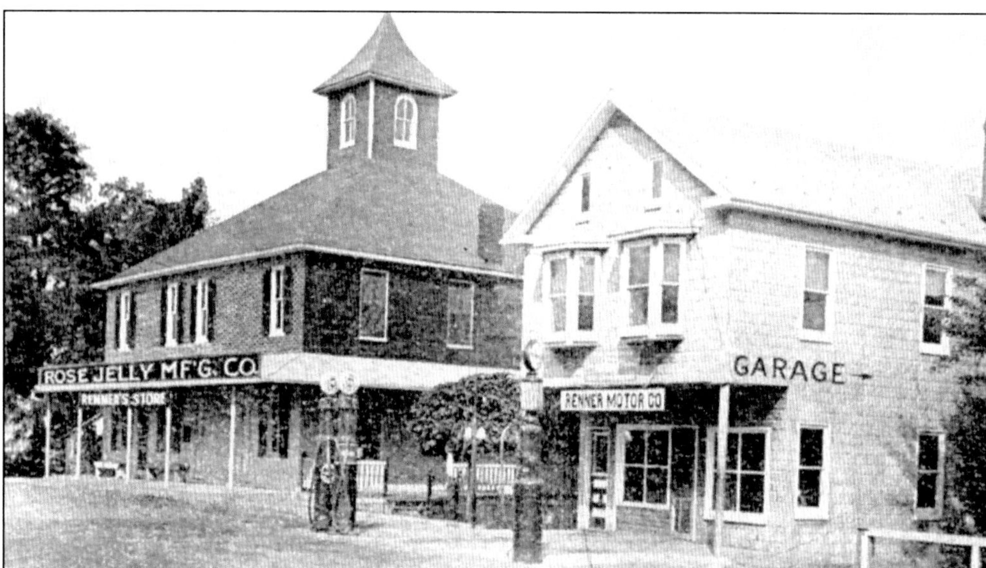

On September 13, 1926, businessmen assembled at Renner's Garage in New Midway to organize a fire department. The first fire company was a small garage built behind Renner's Garage. In 1932, prices were set for fire response: $1 per mile for a chimney fire, $5 to $25 for 2 tanks of water or more, $50 for big fires. Those who could not pay were not charged. (Courtesy Cora Amoss.)

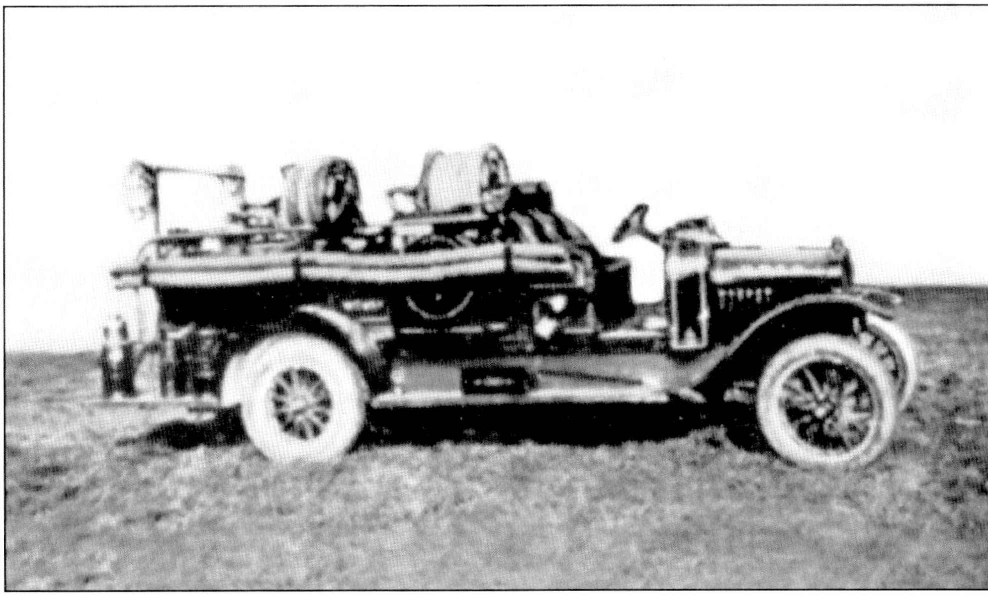

In 1927, the New Midway Volunteer Fire Company purchased its first fire engine, a 1927 Ford Howe, for $2,700. On October 6, 1930, the engine was credited for averting a disaster in Ladiesburg as Chief Jesse Renner and his men made the two-mile response and used chemicals to extinguish a gasoline tank-wagon on fire. The Independent Hose Company of Frederick was called, but the fire was extinguished by New Midway before their arrival. (Courtesy New Midway Volunteer Fire Company [NMFC].)

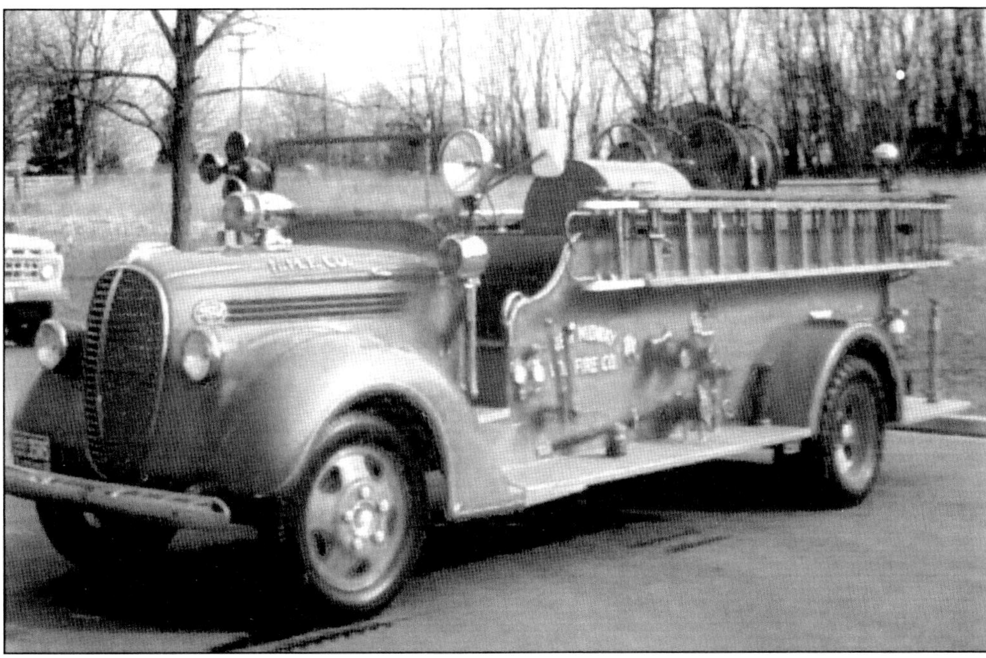

The third fire engine purchased by the New Midway Fire Company was purchased on December 22, 1938. A 1939 Ford Howe engine with a V8 95-horsepower motor was bought for $3,675. A $1,500 trade-in was allowed for the old engine. This unit was the original Engine 91. (Courtesy NMFC.)

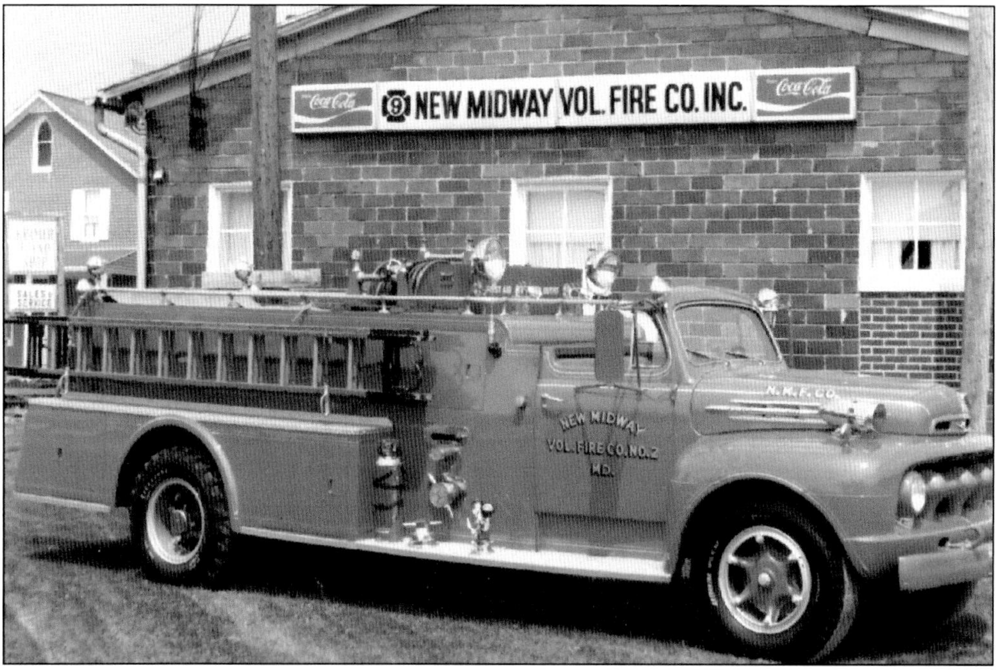

In 1952, the company purchased this Ford Howe pumper. The truck was delivered with a fully enclosed cab, but the membership did not believe it was fair for one driver to have a closed cab while the driver of the 1938 Ford engine was in the open. The members voted to cut the roof off the new engine. The engine was known as Engine 92. (Courtesy NMFC.)

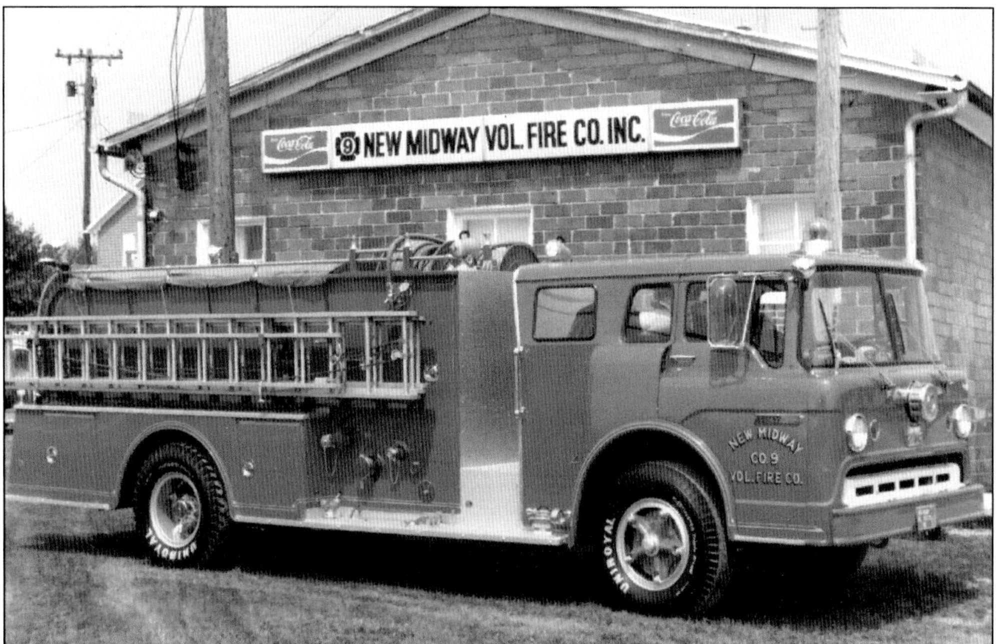

Engine 93 is a 1970 Ford Oren engine purchased in January 1970. The company paid Oren $35,388.20 for the body, and the chassis was purchased from Chesapeake Ford for $9,084. A new Frederick County radio was purchased from Melcom Electronics for $1,224. (Courtesy NMFC.)

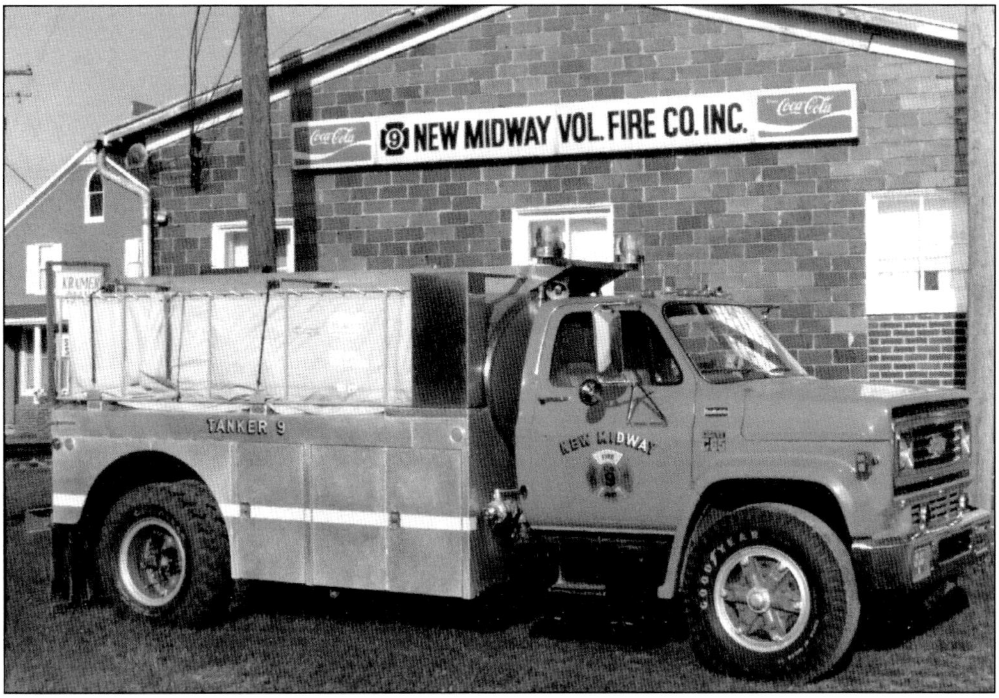

In 1975, a bid of $11,867 was received from Wantz Chevrolet for a Chevrolet chassis for a new tanker. The company approved Heil Company, a milk tanker company of Lancaster, Pennsylvania, to build the body for $8,019. This engine was known as Tanker 9. (Courtesy NMFC.)

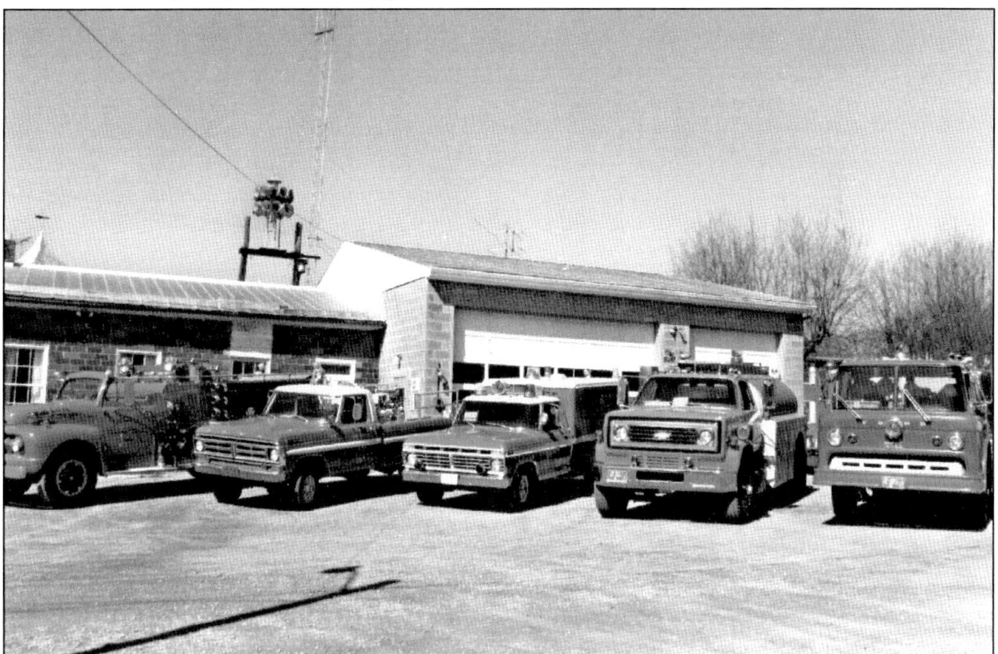
From left to right, Engine 92, Brush 95, Special Unit 9, Tanker 9, and Engine 93 are shown in front of the former fire hall located along the railroad tracks on Route 194 in New Midway. (Courtesy NMFC.)

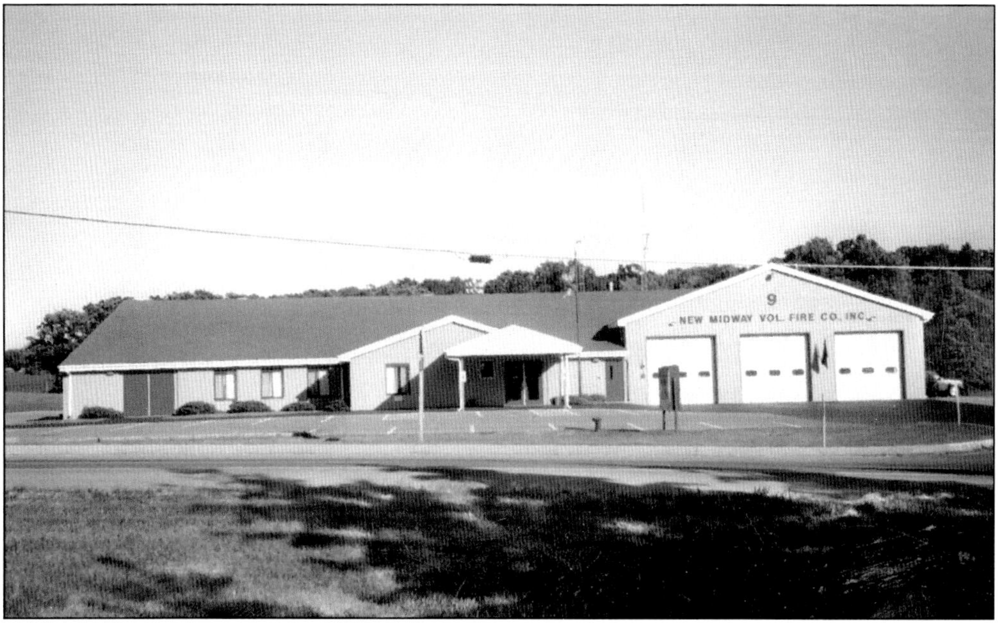

The New Midway Volunteer Fire Company dedicated the new building in 1985. This building is located next to the old firehouse and has a large social hall for fund-raising events as well as a modern engine bay to house the apparatus. (Courtesy NMFC.)

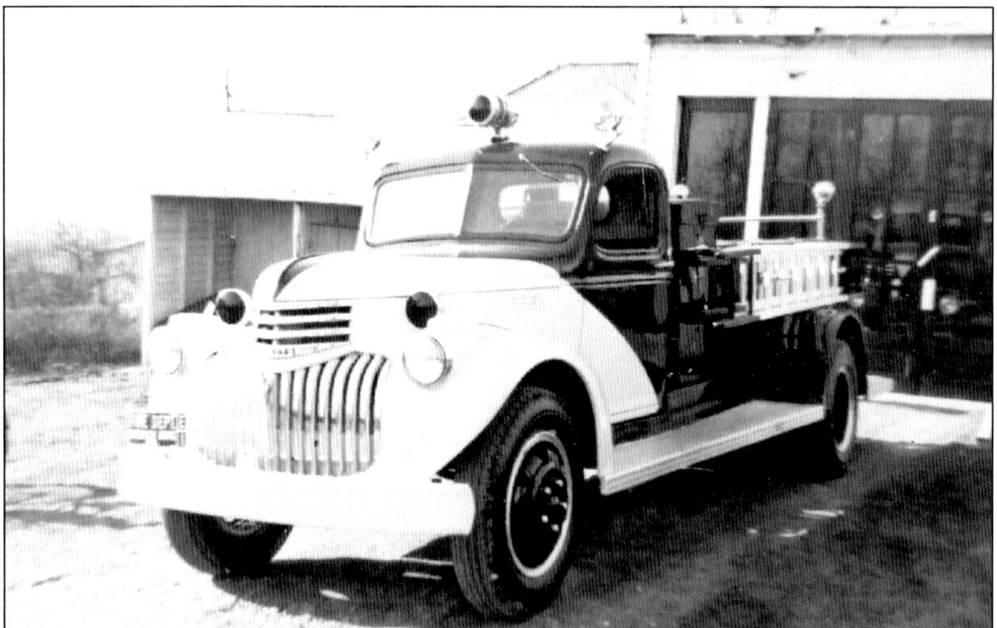

The town of Braddock Heights was founded as a summer resort spot that grew out of the Frederick and Hagerstown Railway trolley. The large frame houses and buildings required local fire protection. The citizens organized a fire company in 1910, many years before the Braddock Heights Fire Company purchased this 1946 Chevrolet engine. This March 1950 picture shows the engine in front of Kern's Garage, where it was kept until the firehouse was built. (Courtesy Braddock Heights Volunteer Fire Company [BHFC].)

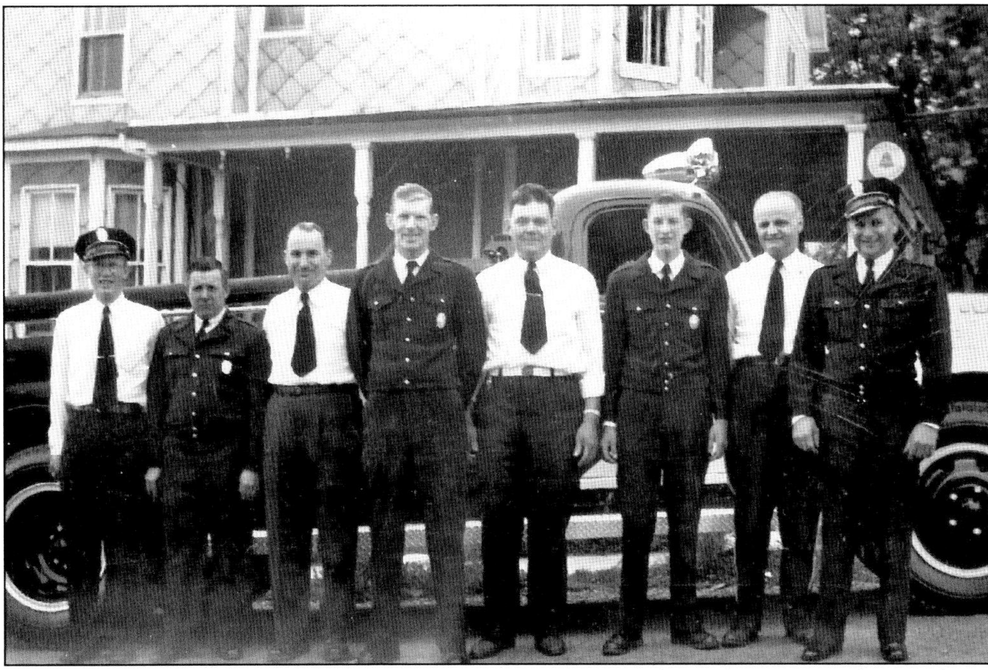

From left to right, Francis Eigenbrode, Austin Cramer, Maynard Wastler, Gene Bishop, Earl Fox, Bubby Cashour, Harold Frushour, and Vernon Cassell Sr. are shown at Kern's Store on Maryland Avenue beside the 1946 Chevrolet engine. (Courtesy BHFC.)

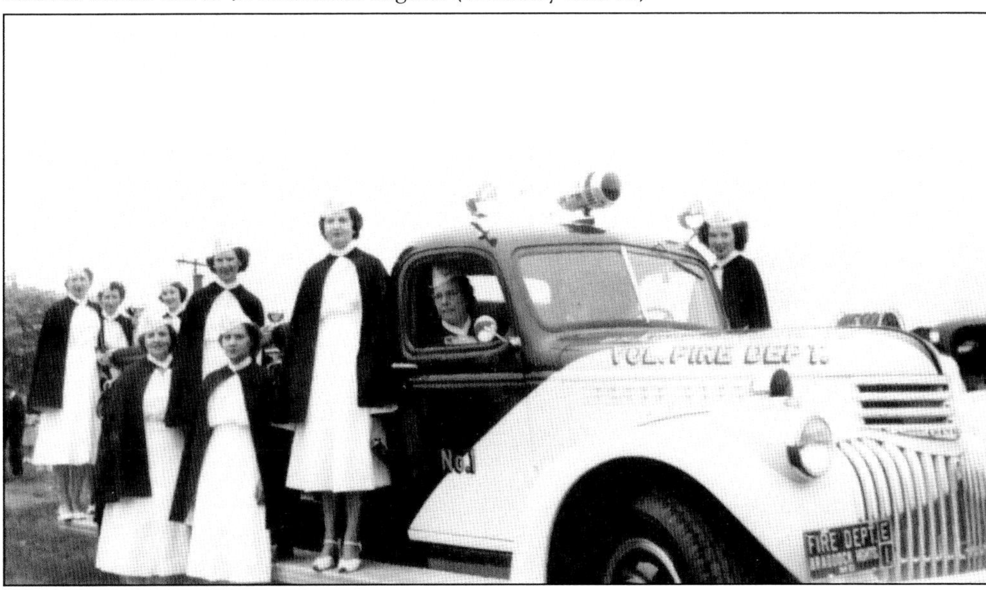

These ladies of the Braddock Heights Volunteer Fire Company were more than just an auxiliary. The women actually responded on calls during times when the men were not available during World War II and the Korean War. At the July 1952 Walkersville parade are, from left to right, (back step) Corrine Wastler, Helen Opel, and Guynetts Neal; (standing front) Kathryn Frushor and Mildred Cassell; (side board) Mary Alexander and Marjorie Wiegand; (cab passenger) Anna Mae Hughes; (driver) Thelma Poole; and (driver side) Pat Bishop. (Courtesy BHFC, donated by Robert and Marjorie Wiegand.)

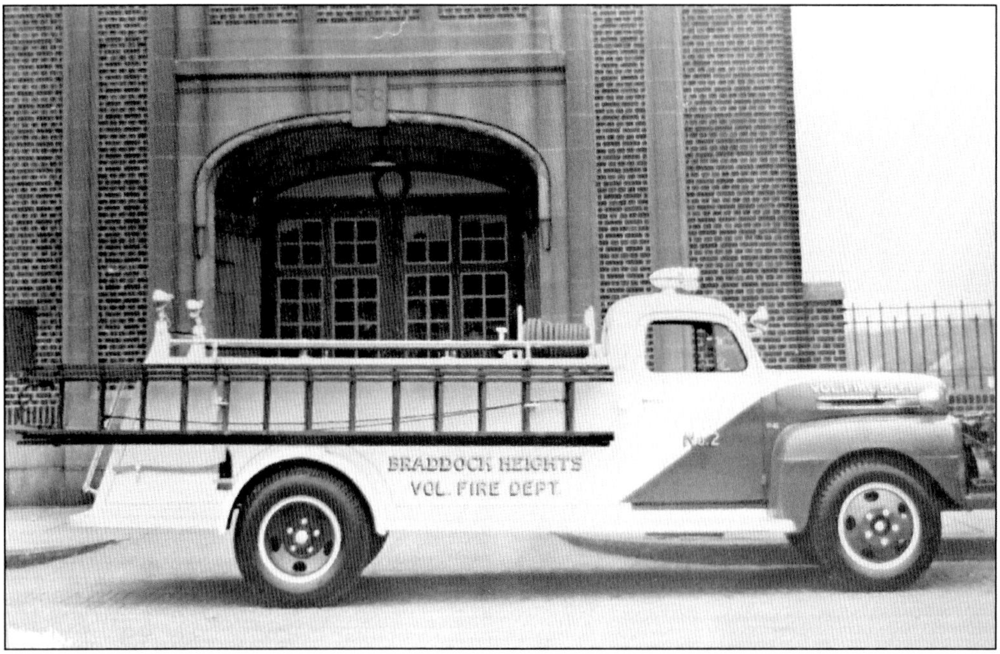

The fire company continued to improve the protection to the community with the purchase of this 1948 Ford, shown prior to delivery in front of the F. L. Anderson Company of Baltimore. This engine had a 500-gallon-per-minute pump and was delivered in April 1949. It served Braddock Heights until it was assigned to the Jefferson substation, where it served until 1976. (Courtesy BHFC, donated by Mildred and Vernon Cassell Sr.)

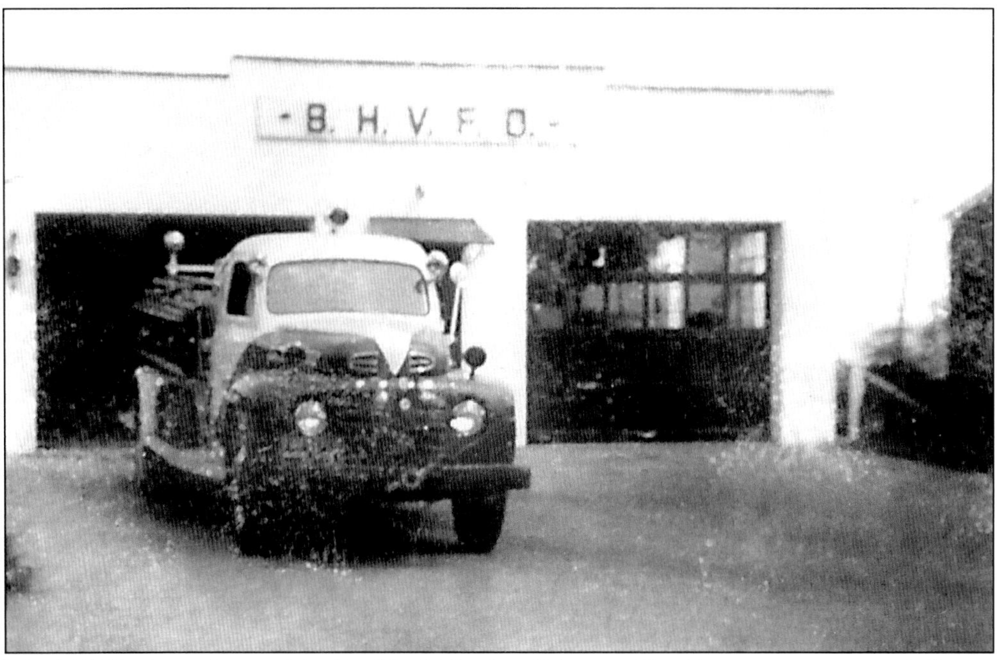

The 1948 Ford is in front of the Braddock Heights Volunteer Fire Company station on Jefferson Boulevard in June 1952. Years later, the doors had to be raised to accommodate larger fire apparatus. (Courtesy BHFC, donated by Genny and Gordon Taylor.)

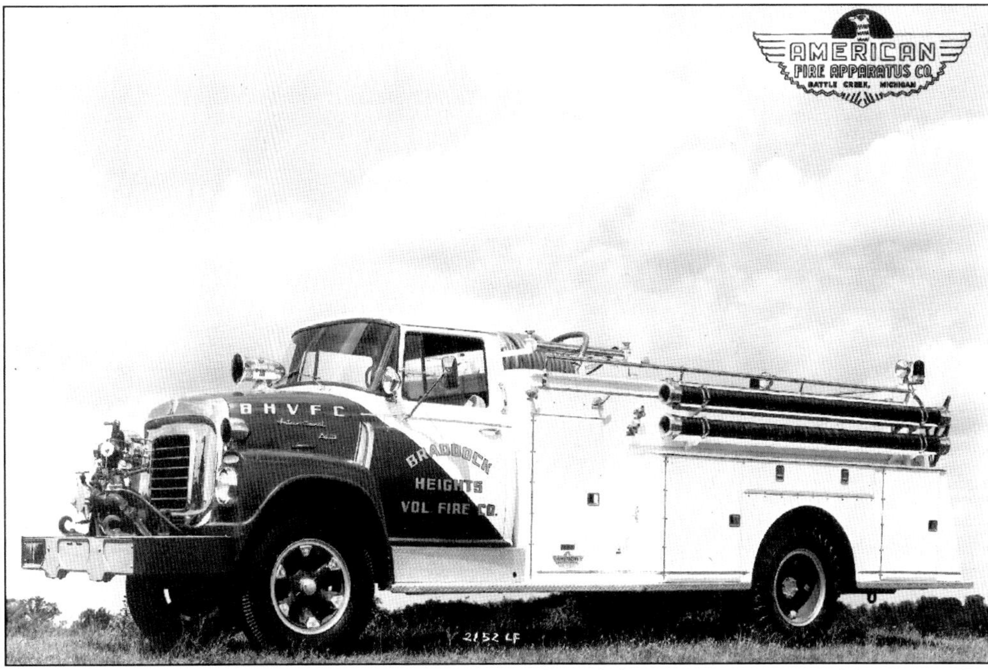

The Braddock Heights Volunteer Fire Company purchased this 1958 International American Fire Apparatus Company engine from Glenn D. Culbert Company, Inc., in College Park. Known as Engine 123, the unit had a 500-gallon-per-minute, front-mounted pump and carried 500 gallons of water. (Courtesy BHFC.)

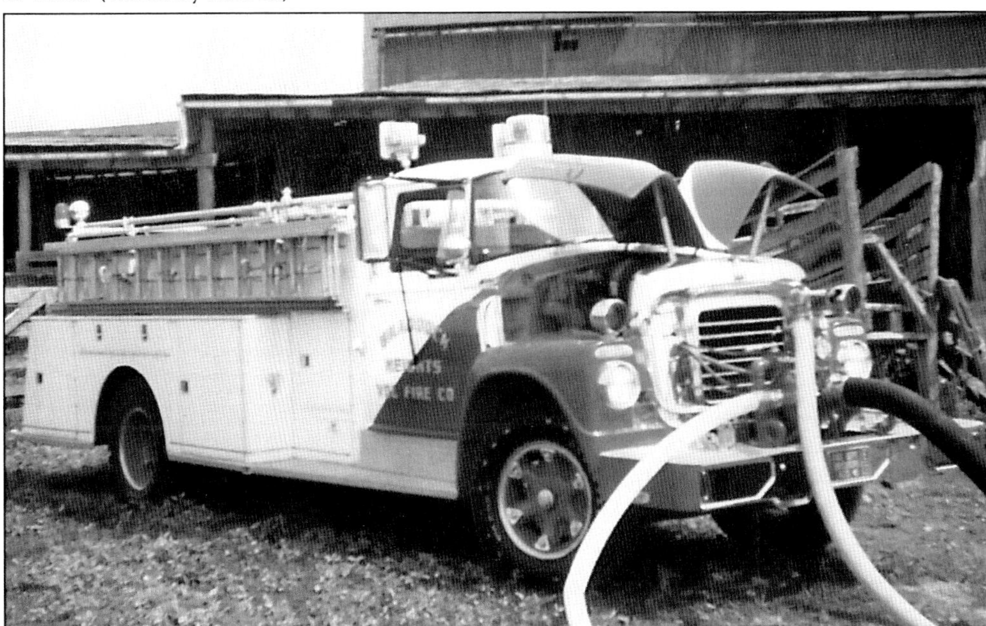

Engine 123 is shown in action at a drill on Bidle Road on Sunday, July 1, 1973. With water flowing, the picture demonstrates the operation of the front-mounted pump. Note the rotating beacon ray on top of the engine, added after the doors were raised on the firehouse. (Courtesy BHFC, donated by Jim Main.)

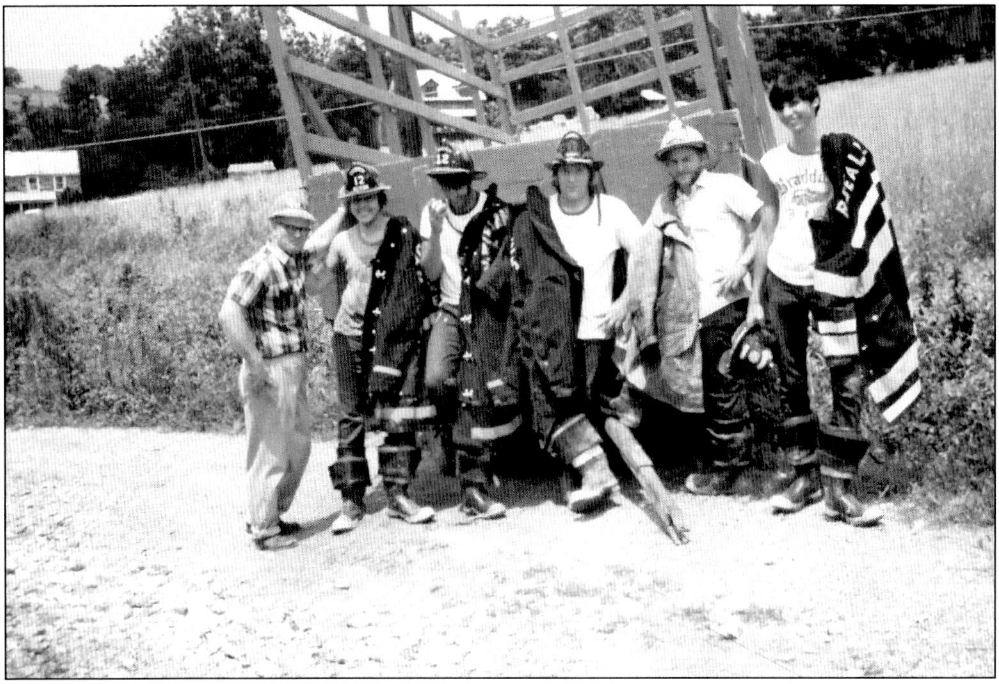

From left to right, Chief Dick Metcalf and firefighters Gary Dorsey, Jim Wiegand, Glenn Smith, Jim Main, and Rick Beall take a break during the Bidle Road drill. (Courtesy BHFC, donated by Jim Main.)

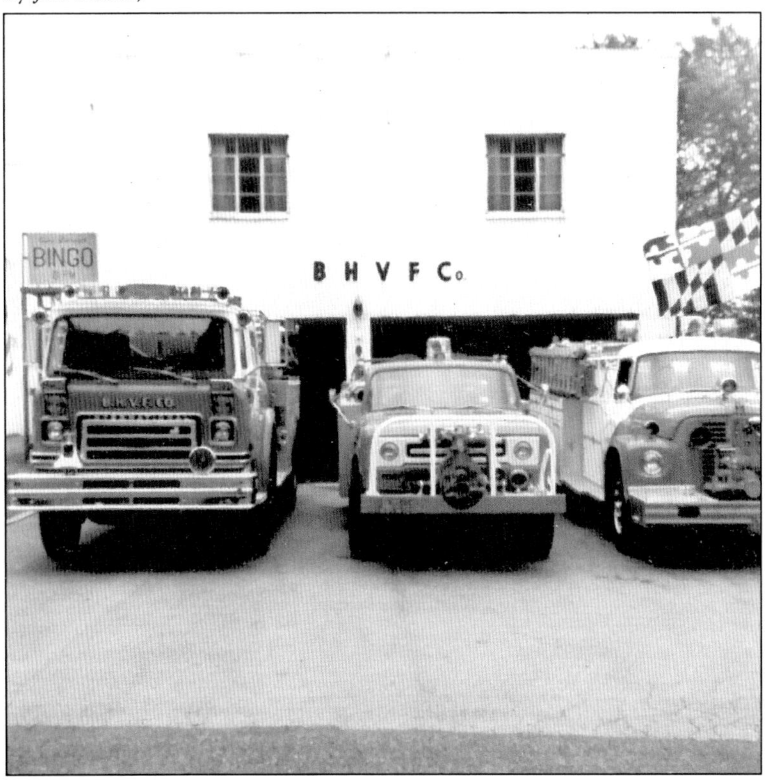

The renovations to the fire hall included raising the doors and adding a second floor. Units shown from left to right are Engine 122, a 1975 International Pierce engine; Brush 125, a 1970 International brush truck; and Engine 121, a 1966 International. (Courtesy BHFC, donated by Betty and Jim Wiegand.)

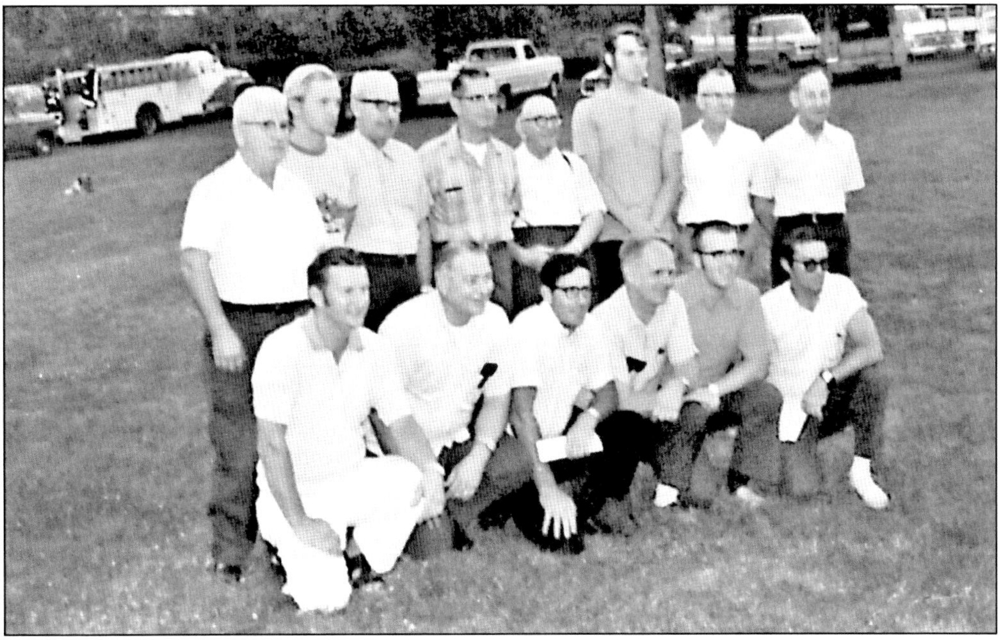

The 1972–1973 fire company officers attending the July membership chicken barbecue at the Braddock Pool are, from left to right, (first row) Vernon Cassell Jr., chief; Gordon Taylor, captain; William "Sam" Lenhart, captain; Richard Metcalf, captain; William Metcalf, lieutenant; and Doug Abrecht, lieutenant; (second row) Herman Toms, vice president; James Main, director; Maynard Wastler, director; Horace Opel, director; Fulton Nicodemus, director; Rick Opel, secretary; Arthur Throne, treasurer; and Grafton Cost, president. (Courtesy BHFC.)

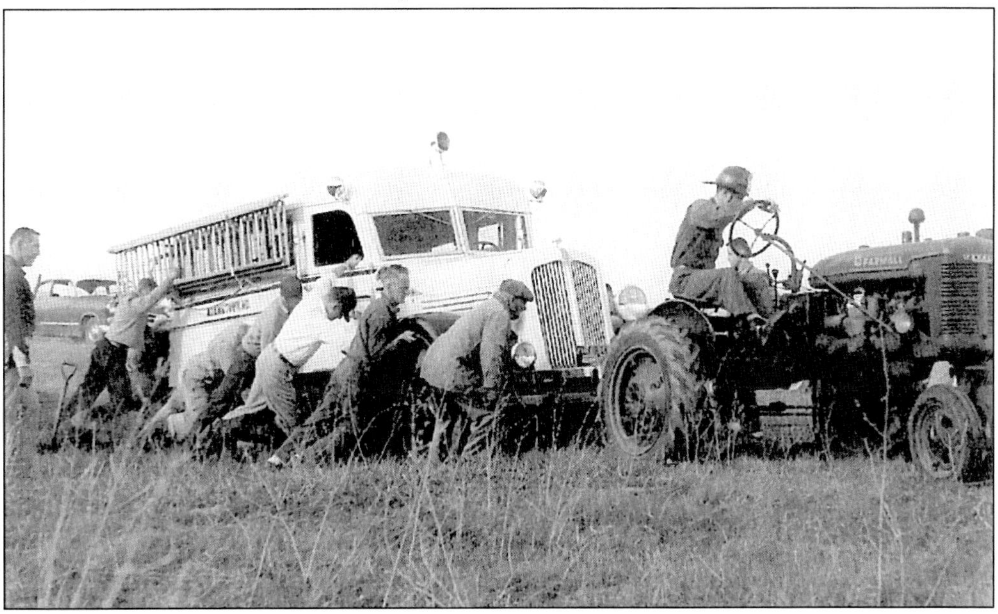

A member of Carroll Manor Volunteer Fire Company had to get his tractor to pull the 1937 Sedan Seagrave out of a muddy field after it was used on a fire. Extra manpower was needed to help the tractor pull the engine. This engine was purchased from the Independent Hose Company No. 1 in Frederick. (Courtesy Carroll Manor Fire Company [CMFC].)

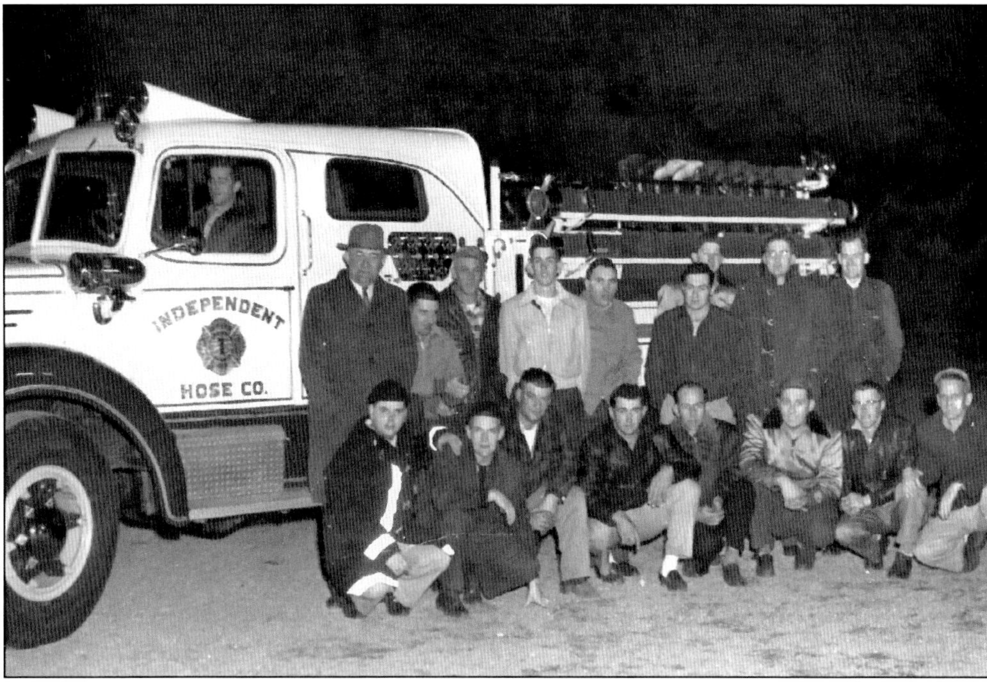

Members of the Carroll Manor Fire Company who attended the first basic firefighting class are shown next the Independent Hose Company engine. Members of the Independent Hose Company helped Carroll Manor organize and assisted with training. Frank Abrecht of the Independents is driving the Mack engine. (Courtesy CMFC.)

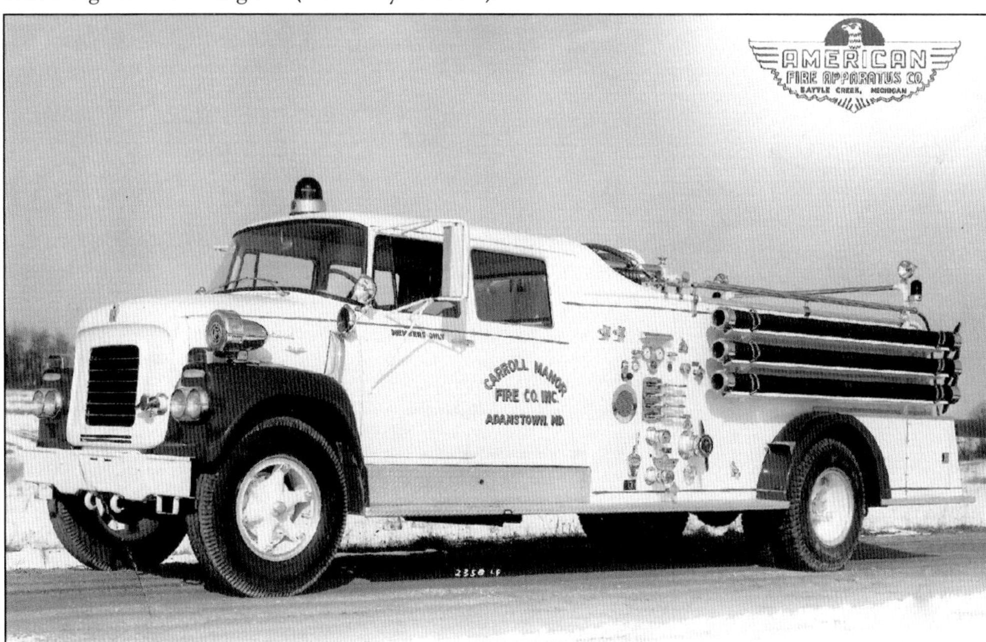

The Carroll Manor Fire Company adopted green and white as company colors in this American Fire Apparatus Company shot of Engine 141, a 1961 International pumper. This unit had a four-door cab, a safety advancement ahead of its time in 1961. (Courtesy CMFC.)

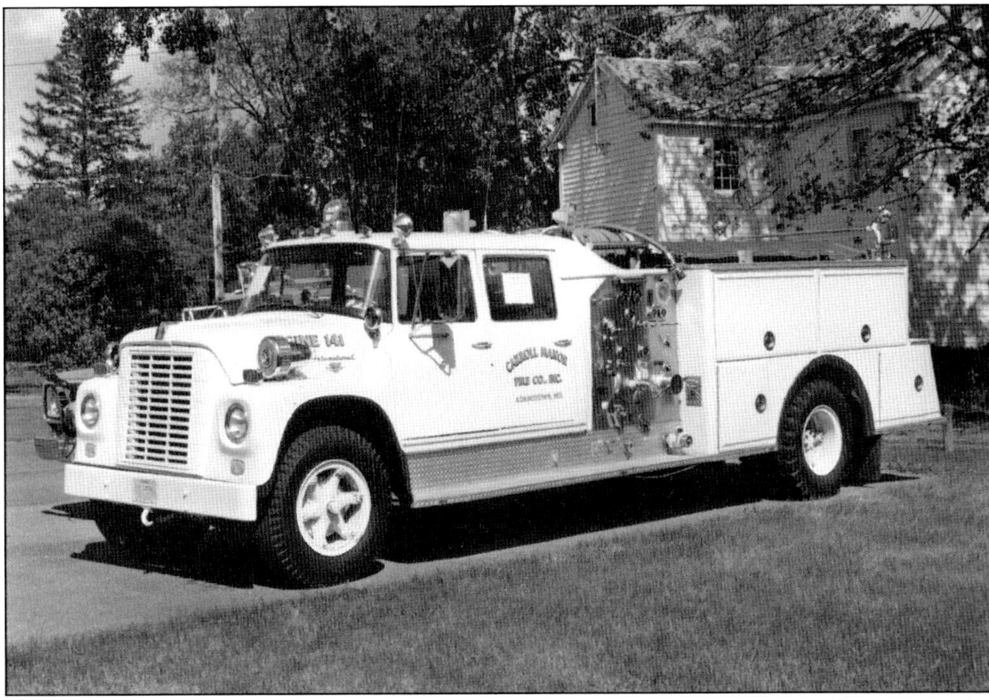

Nearly a twin to the earlier sister engine, Engine 142 was a 1968 International engine with a four-door cab for the safety of the firefighters. (Courtesy CMFC.)

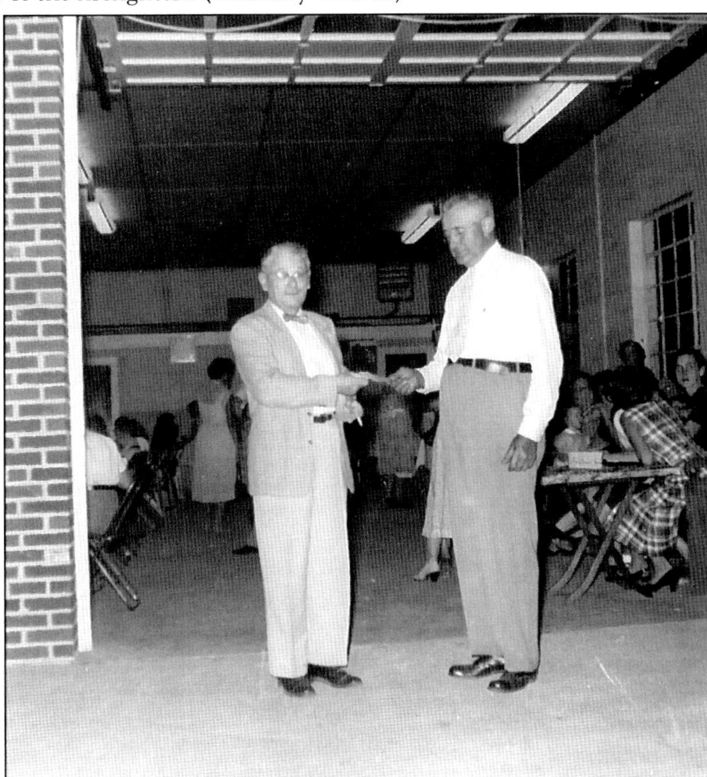

Fund-raising is the backbone of a volunteer fire company. The Carroll Manor Volunteer Fire Company has always held activities to raise money for the department. Unidentified members are working at a bingo held on February 5, 1955. (Courtesy CMFC.)

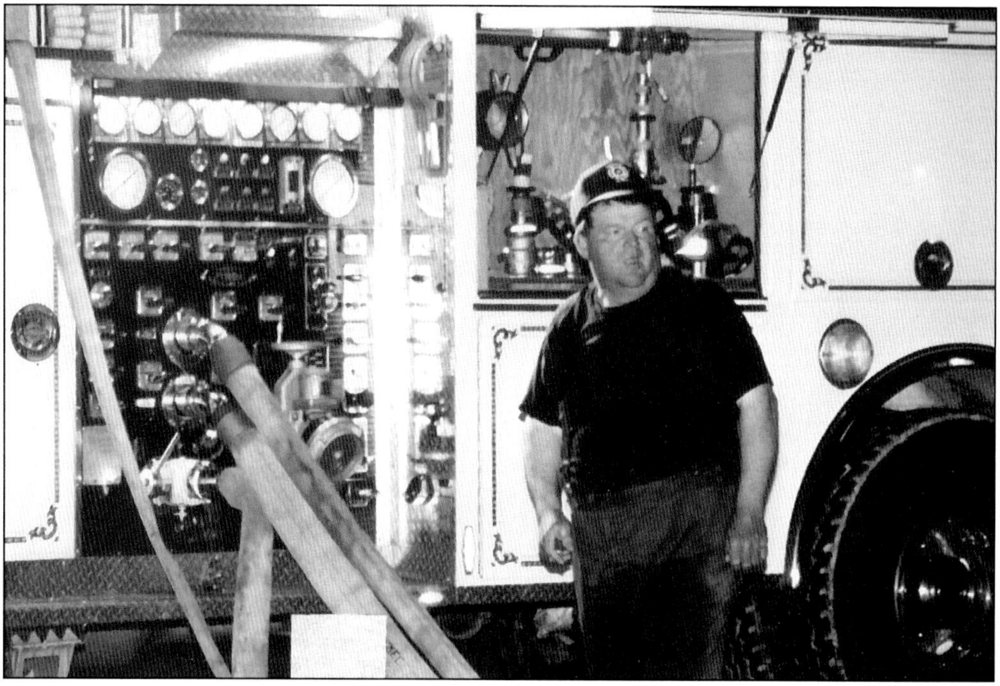

Shown operating the pumps during a working fire, John Mills was a dedicated member, a dependable driver, past chief, and true friend to the fire company. He spent his life to better the fire company and his community. (Courtesy CMFC.)

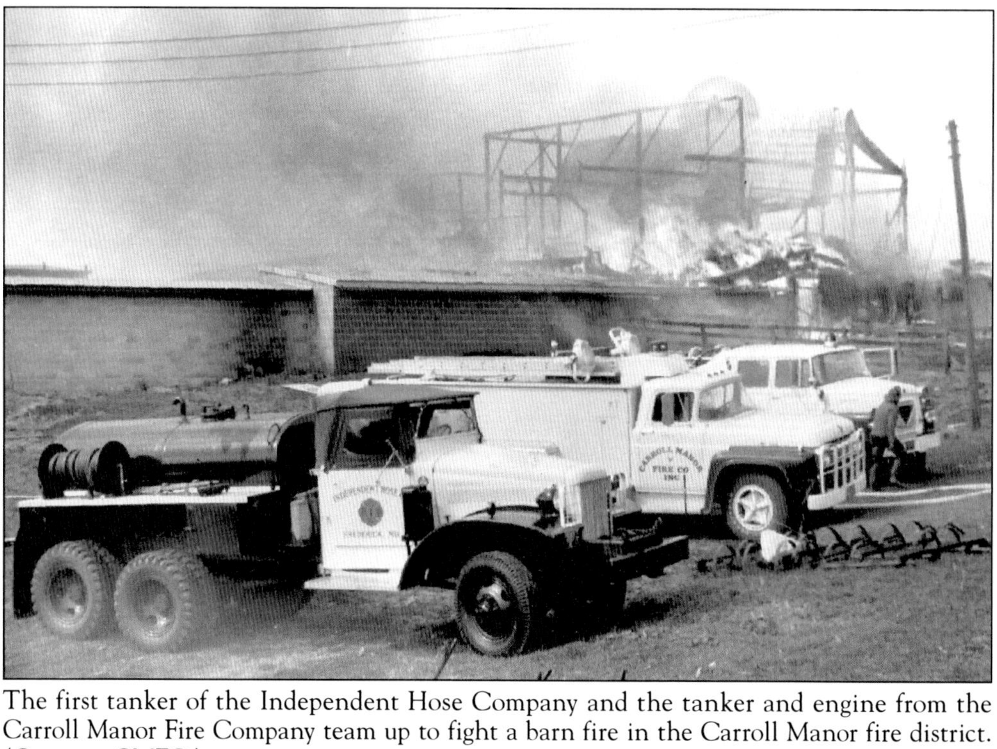

The first tanker of the Independent Hose Company and the tanker and engine from the Carroll Manor Fire Company team up to fight a barn fire in the Carroll Manor fire district. (Courtesy CMFC.)

Carroll Manor to the rescue as this black lab is carried to safety by two Carroll Manor firefighters. Former fire chief Max Keyser is on the right; the firefighter on the left is unidentified. (Courtesy Susie Kyle.)

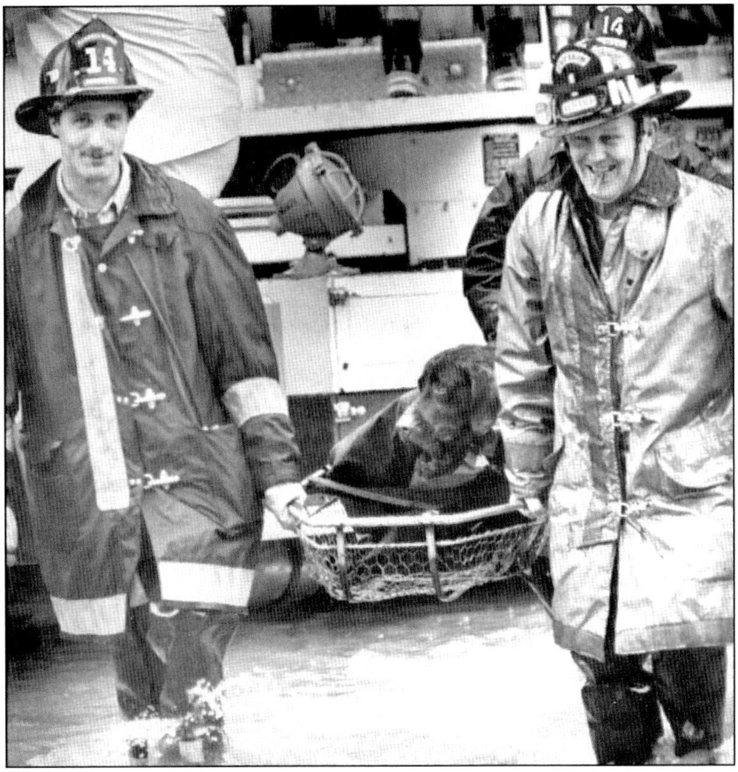

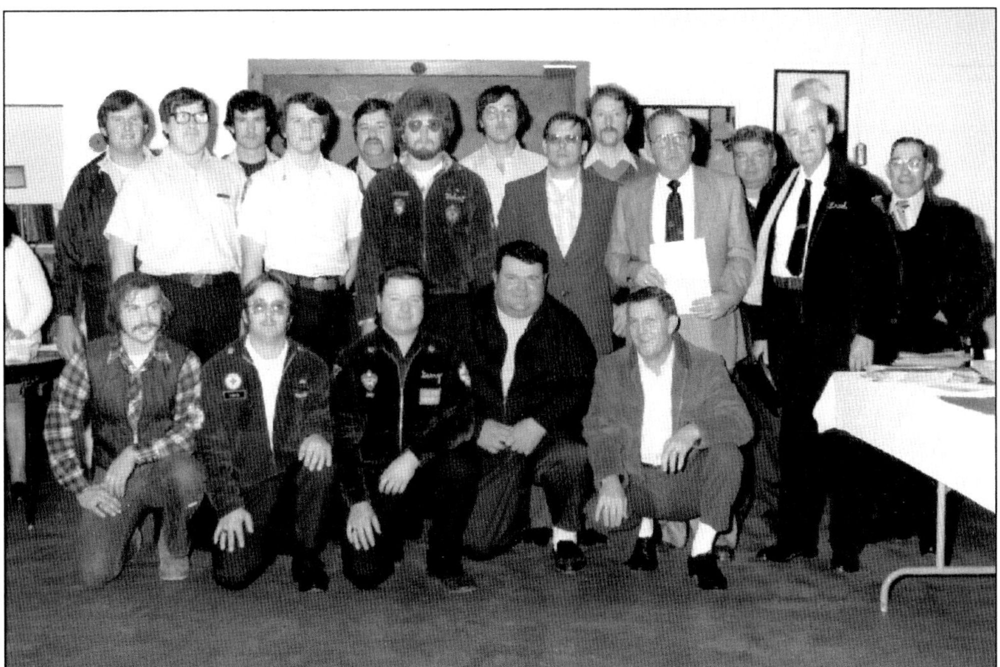

The officers of the Carroll Manor Fire Company are shown after a company meeting at the Adamstown fire station. The member standing holding the paper was L. Eugene Johnson, the longtime president of the company and dedicated member of the ambulance crew. (Courtesy CMFC.)

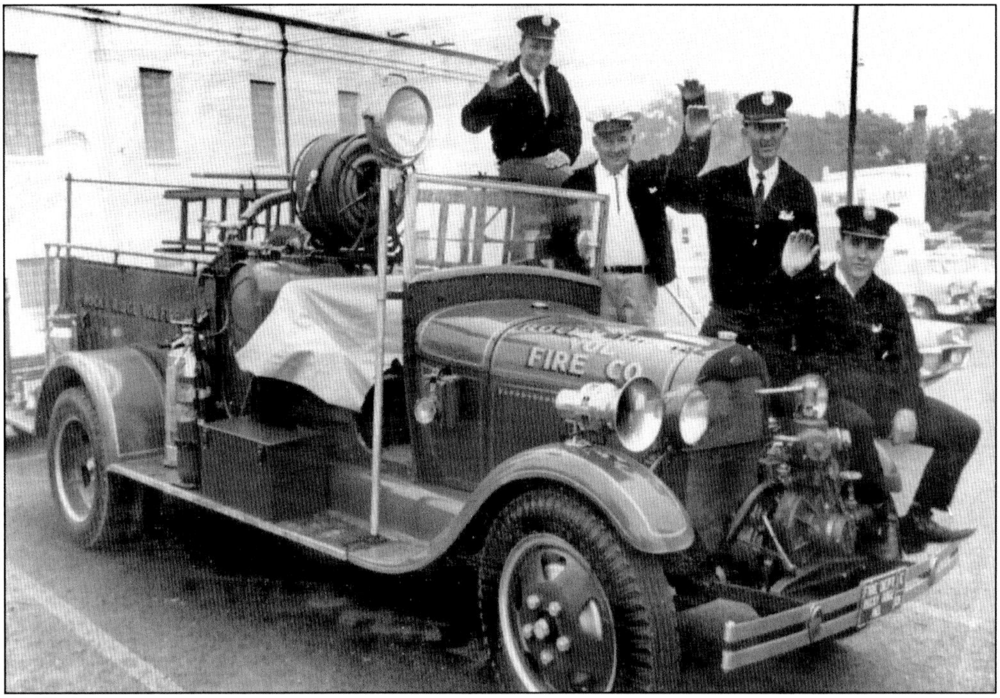

From left to right, Dennis Mathias, Albert Tabler, Robert Albaugh, and Robert Burrier have just arrived back in Rocky Ridge after a trip to the Maryland State Firemen's Association convention in Ocean City in 1971. The 1929 Ford engine was originally owned by the Vigilant Hose Company of Emmitsburg. (Courtesy Rocky Ridge Volunteer Fire Company [RRFC].)

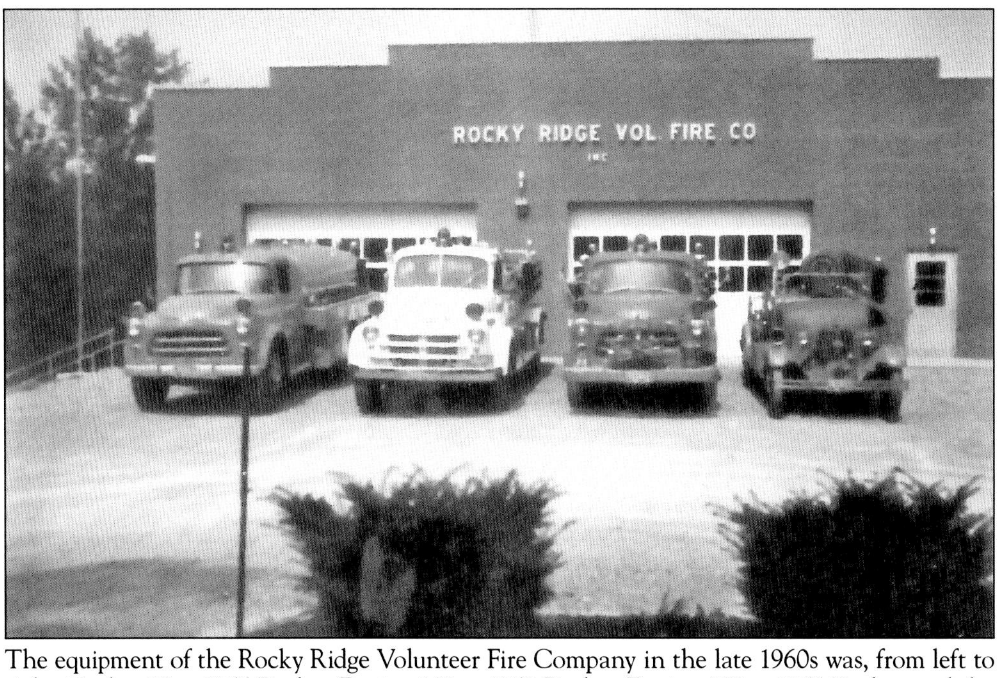

The equipment of the Rocky Ridge Volunteer Fire Company in the late 1960s was, from left to right, Tanker 13, a 1955 Dodge; Engine 133, a 1950 Dodge; Engine 132, a 1955 Dodge; and the 1929 Model A Ford. (Courtesy RRFC.)

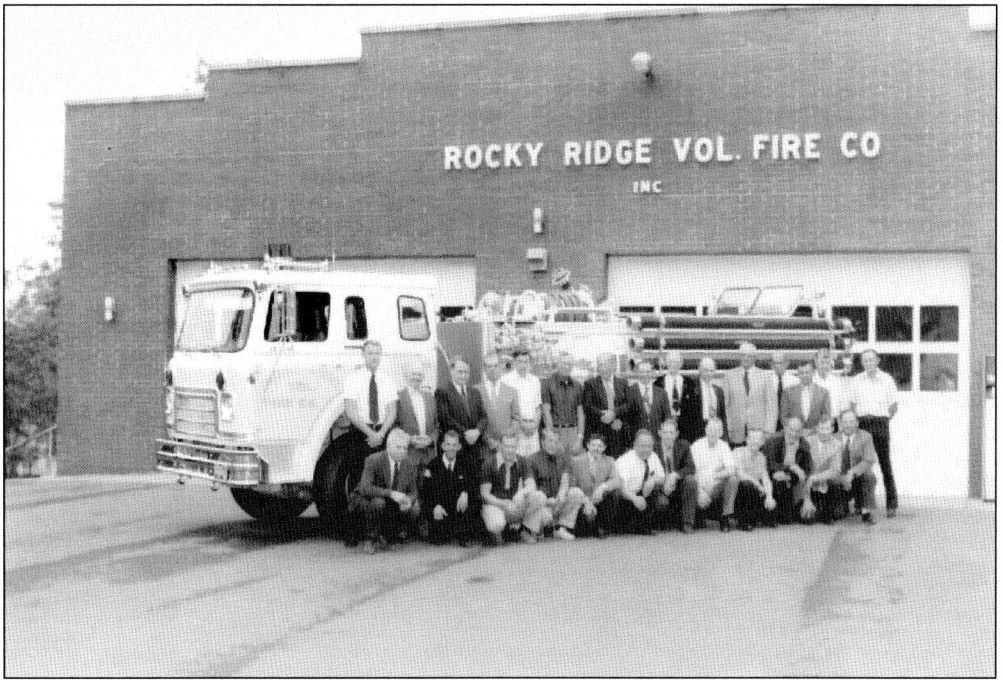

Members of the Rocky Ridge Volunteer Fire Company pose in front of Engine 134, a 1974 International engine. In later years, this engine was rebuilt to include a 1,500-gallon water tank and responded as Engine/Tanker 134. (Courtesy RRFC.)

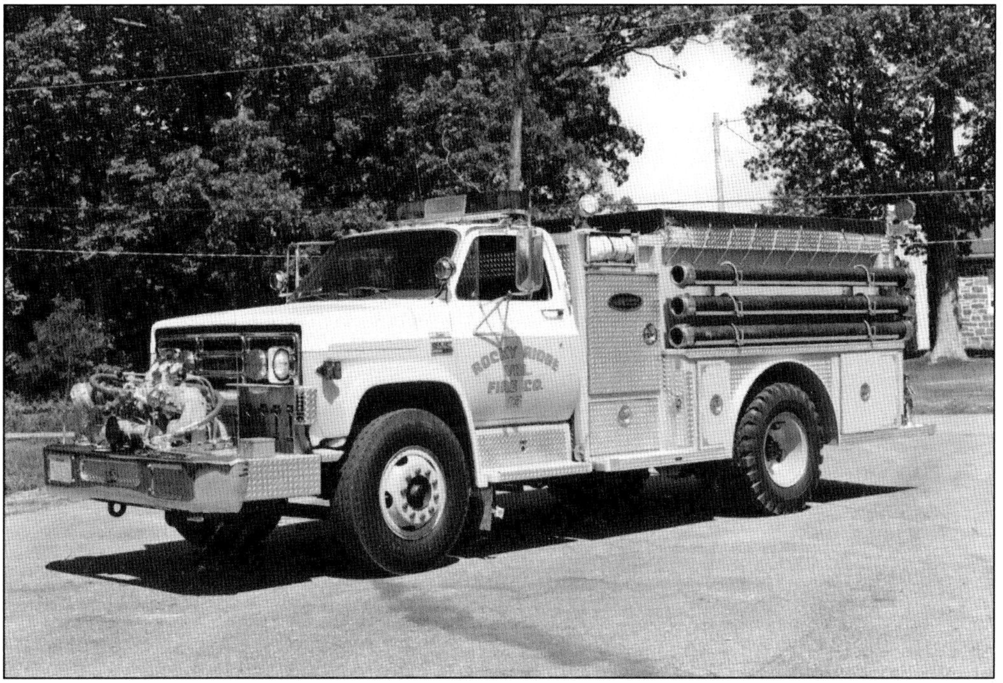

The last front-mounted pump still in service in Frederick is Engine 133, a 1978 GMC E-One front-mounted pumper. The engine has a 750-gallon-per-minute pump and carries 500 gallons of water. (Courtesy Warren Jenkins.)

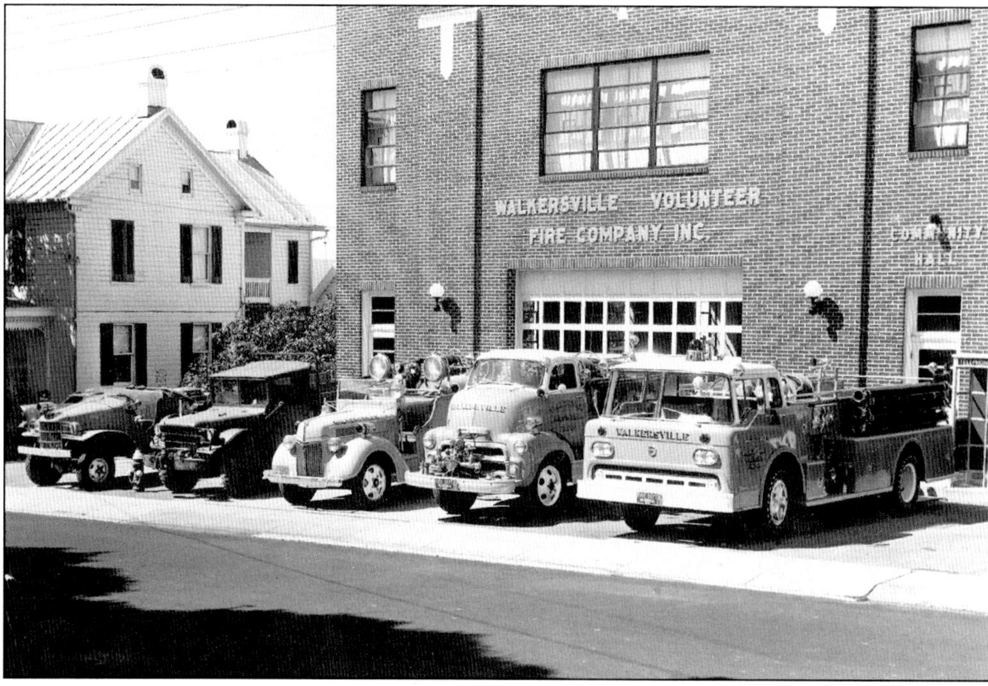

This 1958 picture in front of the old Walkersville fire hall shows, from left to right, two army-surplus vehicles used as brush trucks, 1940 Ford Howe, a unique 1952 Chevrolet with front-mounted pump, and the "new" 1958 Ford Pirsch. (Courtesy Walkersville Volunteer Fire Company [WVFC].)

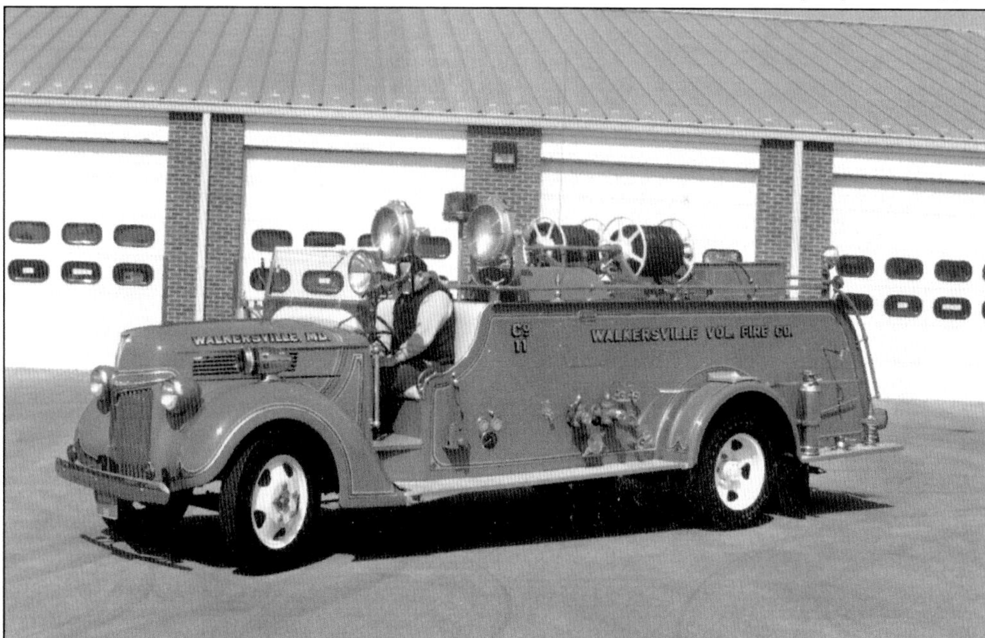

The first engine purchased by Walkersville was Engine 113, a 1940 Ford Howe engine with a 350-gallon-per-minute pump and 200 gallons of water. Walkersville is noted for the color of the apparatus: all equipment is painted a bright "Walkersville" blue. The engine is in front of the new firehouse. (Courtesy Warren Jenkins.)

Old Engine 111, a 1958 Ford Pirsch, is still seen in the Frederick area at local parades. Until recently, the engine was owned by Reggie Kephart of Johnsville. The truck has recently been sold to a collector in Virginia, who drove it all the way to Walkersville for the first fire company parade in 13 years. (Courtesy BHFC, donated by Jim Wiegand.)

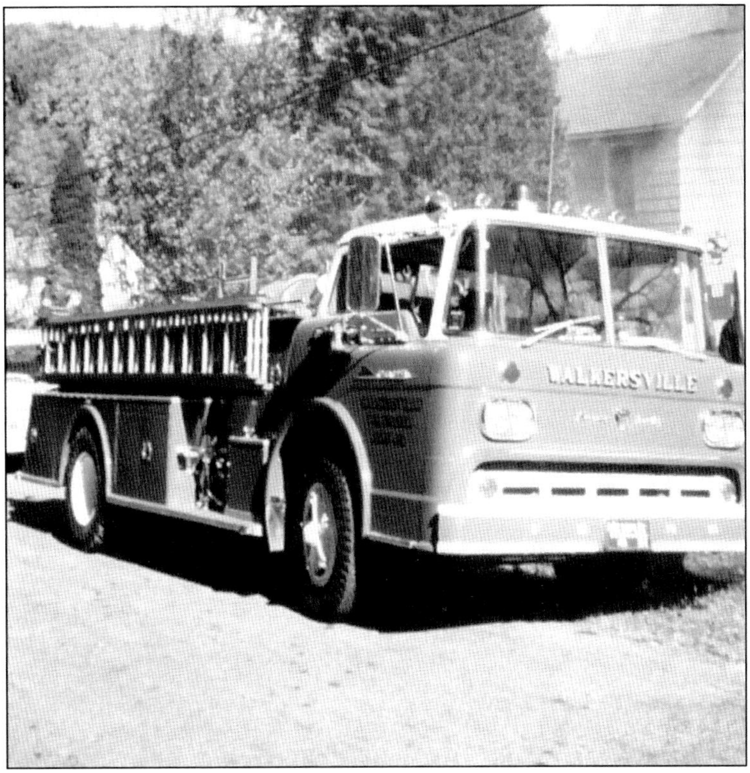

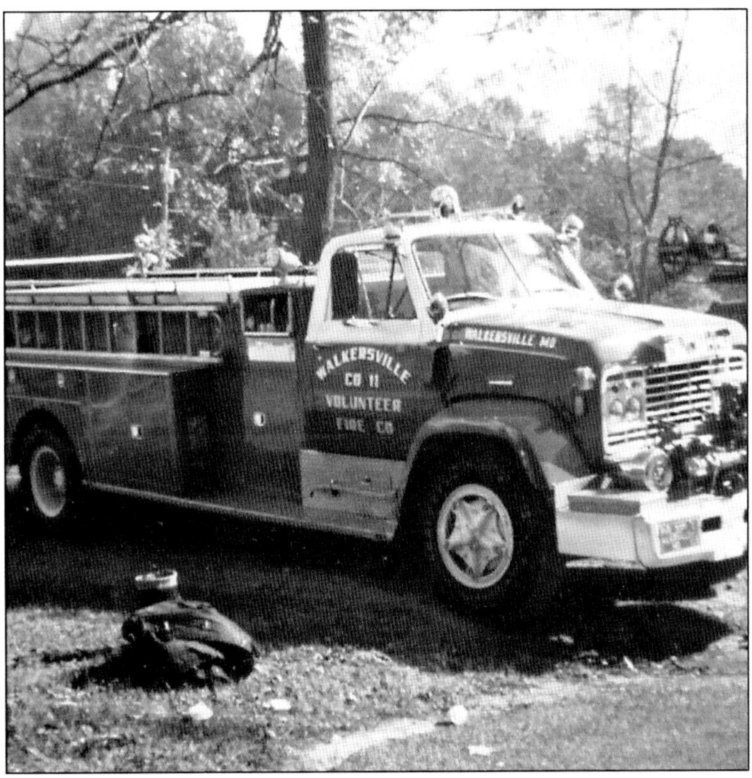

This 1970 front-mounted GMC pumper was built by American Fire Apparatus and carried 1,000 gallons of water. It was later replaced by a Pierce engine tanker with 1,500 gallons of water. (Courtesy BHFC, donated by Jim Wiegand.)

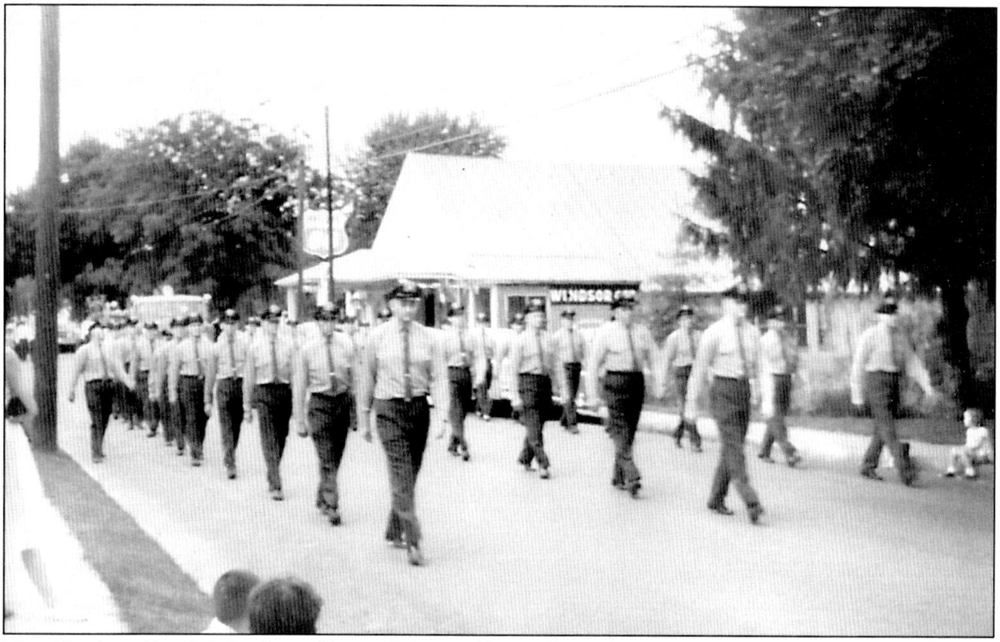

The Uniform Rank Marching Unit of the Walkersville Volunteer Fire Company won first place at the Woodsboro Firemen's Parade on June 26, 1962. The marching unit had 40 members in formation for a very impressive showing. (Courtesy WVFC.)

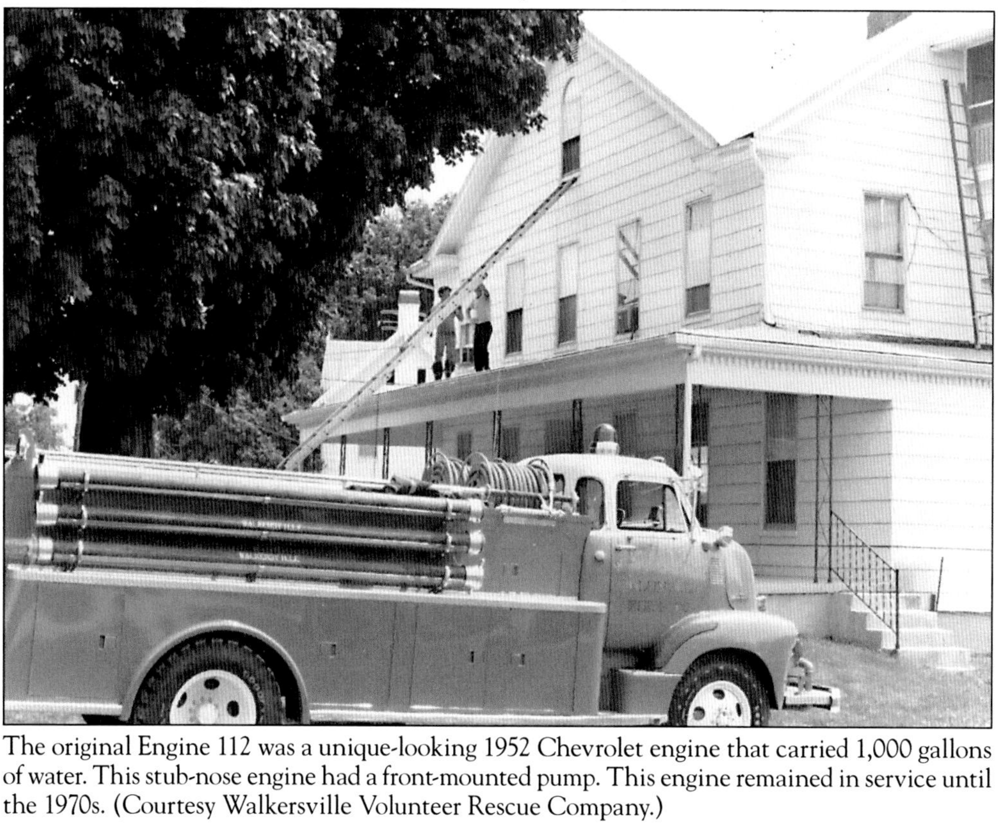

The original Engine 112 was a unique-looking 1952 Chevrolet engine that carried 1,000 gallons of water. This stub-nose engine had a front-mounted pump. This engine remained in service until the 1970s. (Courtesy Walkersville Volunteer Rescue Company.)

Three

THE COUNTY GROWS
1950 to 2005

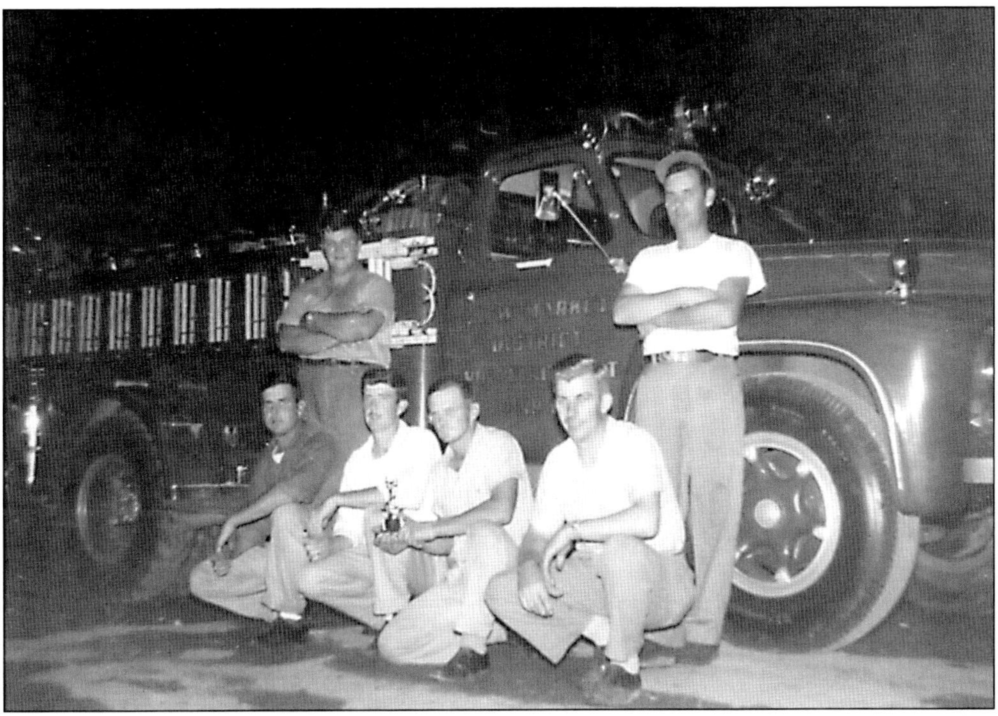

Engine 151, a 1955 Ford American pumper, was the first engine purchased by the New Market District Volunteer Fire Department. The members in this October 1959 picture, from left to right, are (kneeling) Mike Wilcom, Jerome Wilcom, Tom Maher, and Lucian Falconer; (standing) Gene Zimmerman and Bill Clark. (Courtesy New Market District Volunteer Fire Department [NMVFD].)

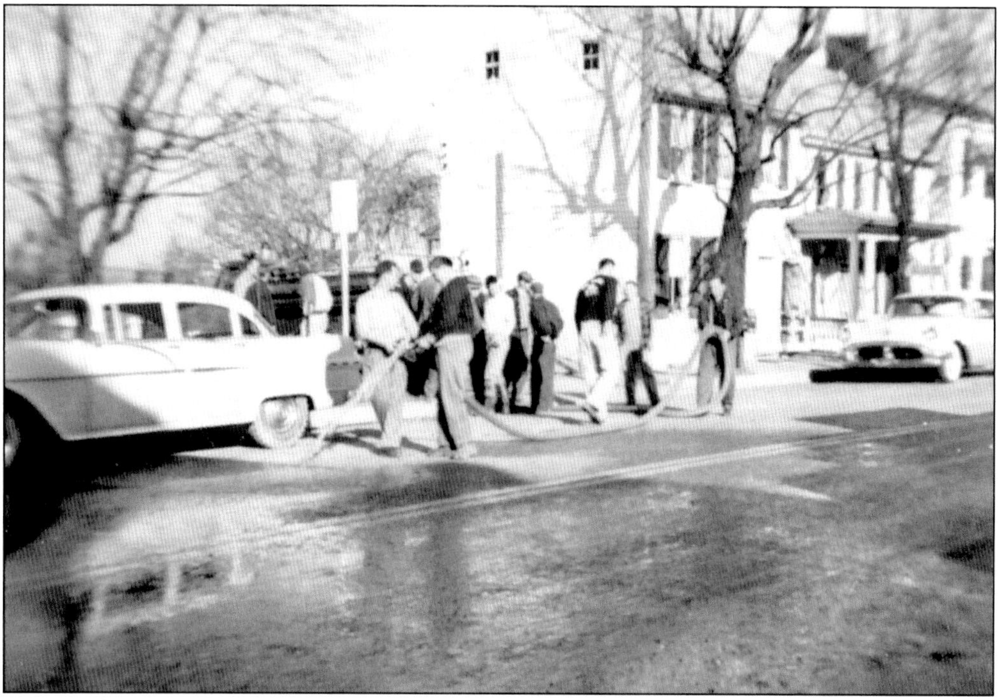

In December 1957, Bill Clark, Tom Maher, Gene Zimmerman, Lucian Faulkner, and Frank Smith are pictured from left to right working with other members of the community to test the new fire pump to supply the hydrant in the center of town. (Courtesy NMVFD.)

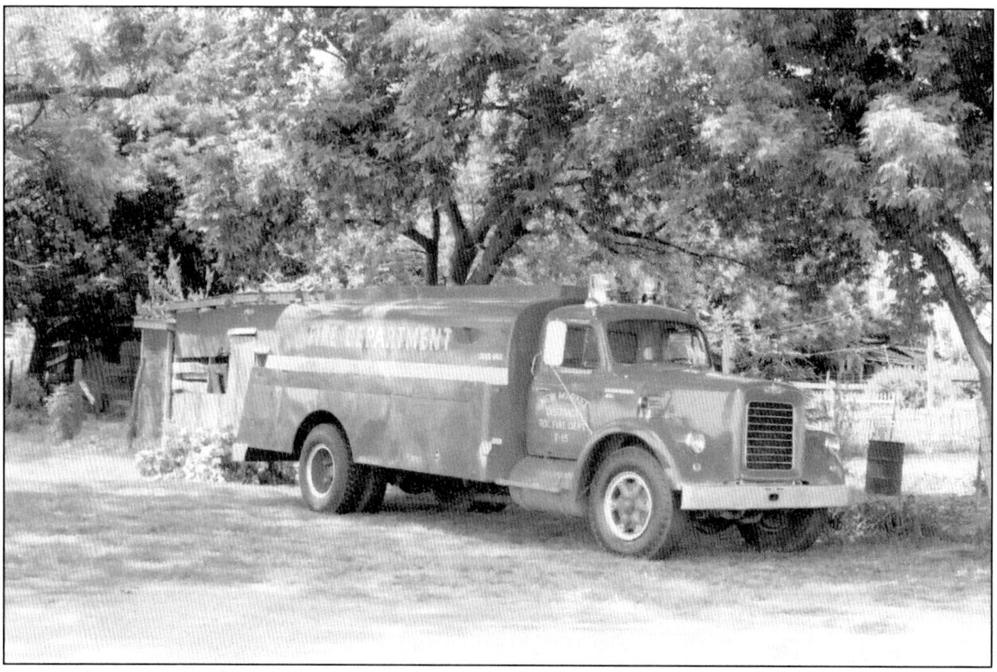

To supply water in the rural areas of eastern Frederick County, the company purchased this International tanker. The unit carried over 2,000 gallons of water and was the largest tanker in Frederick County for many years. The unit was designated Tanker 15. (Courtesy NMVFD.)

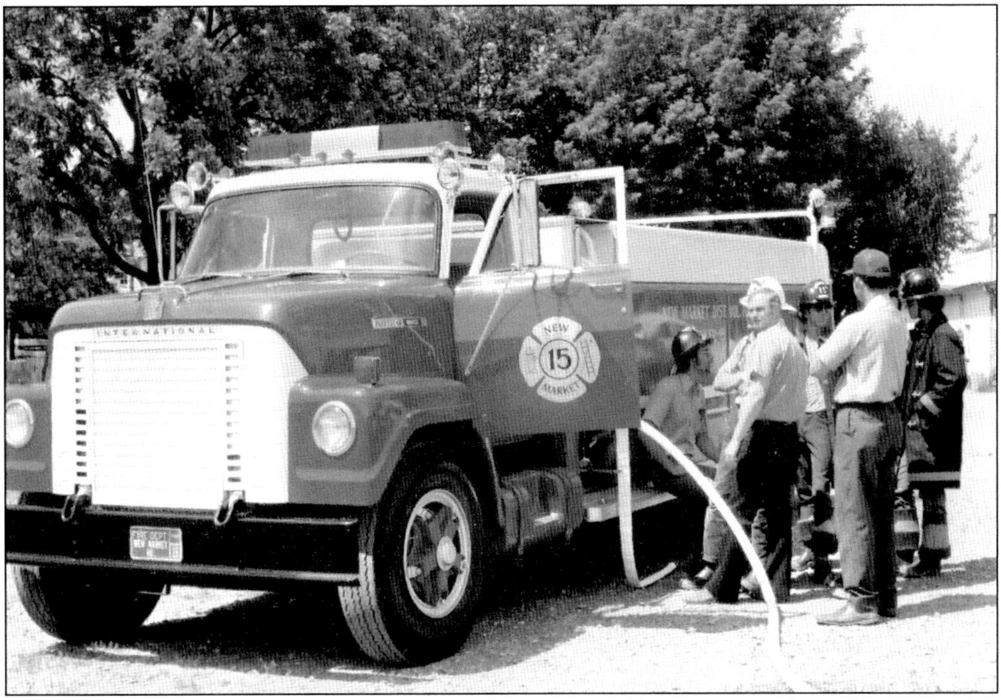

The original tanker was replaced with this 1972 International/Firemaster tanker. The new Tanker 15 carried 2,200 gallons of water and had a 450-gallon-per-minute pump. It remained in service until 1993. (Courtesy NMVFD.)

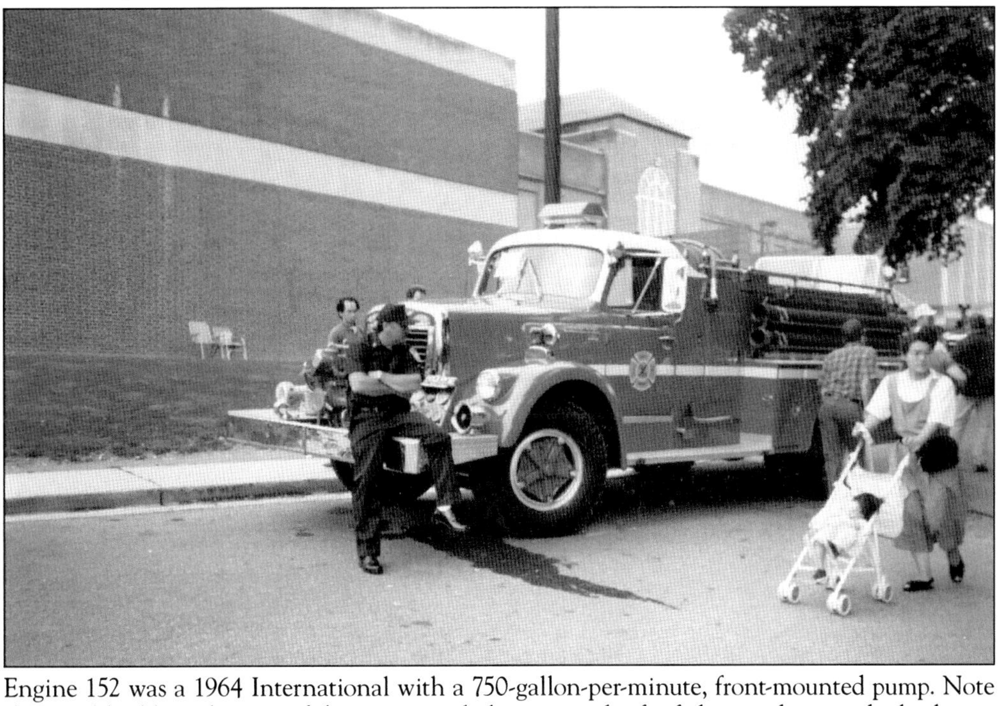

Engine 152 was a 1964 International with a 750-gallon-per-minute, front-mounted pump. Note the windshield on the rear of the engine to help protect the firefighters riding on the back step from being hit by road debris. (Courtesy NMVFD.)

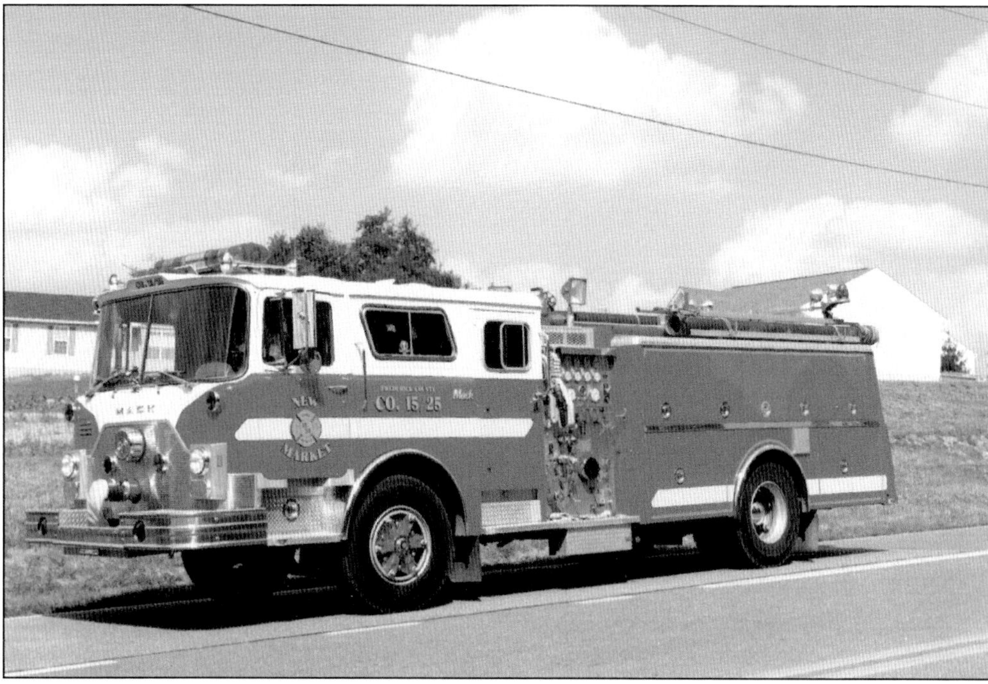

Engine 153 was a 1984 CF Mack with a fully enclosed cab. This unit eventually saw service in the Green Valley substation and ran as Engine 252 until 2003. (Courtesy Warren Jenkins.)

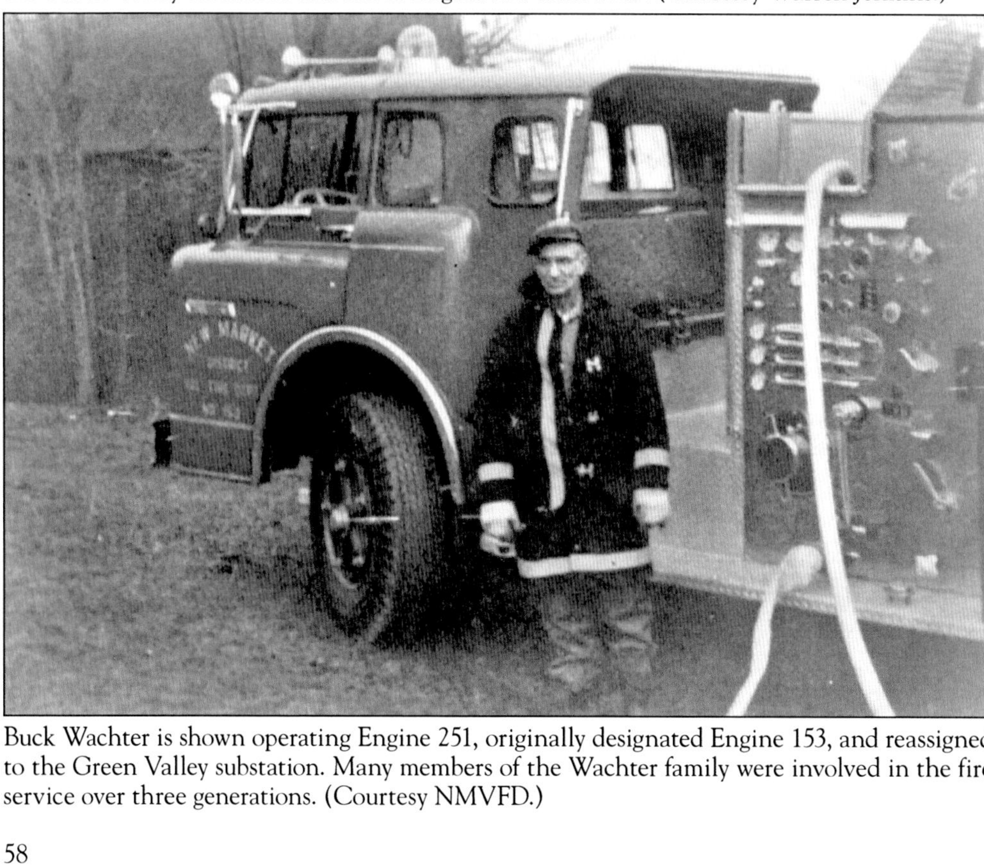

Buck Wachter is shown operating Engine 251, originally designated Engine 153, and reassigned to the Green Valley substation. Many members of the Wachter family were involved in the fire service over three generations. (Courtesy NMVFD.)

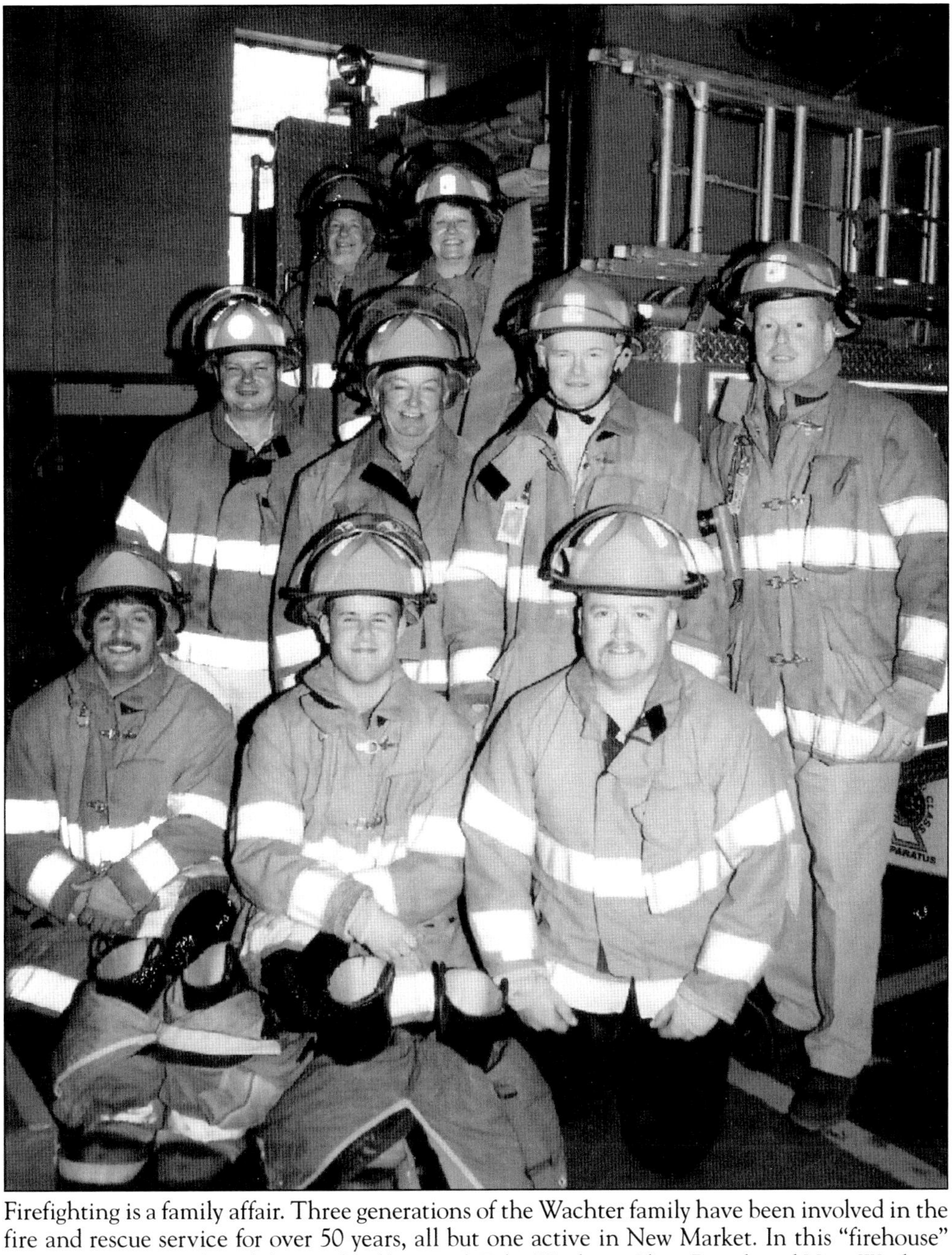

Firefighting is a family affair. Three generations of the Wachter family have been involved in the fire and rescue service for over 50 years, all but one active in New Market. In this "firehouse" family portrait are, from left to right, (first row) Zeke Wachter, Chris Beard, and Vern Wachter; (second row) Rick Wachter, Dick Wachter, Donald Wachter, and Bruce Wachter; (third row) Doc Wachter and Shirley Beard.

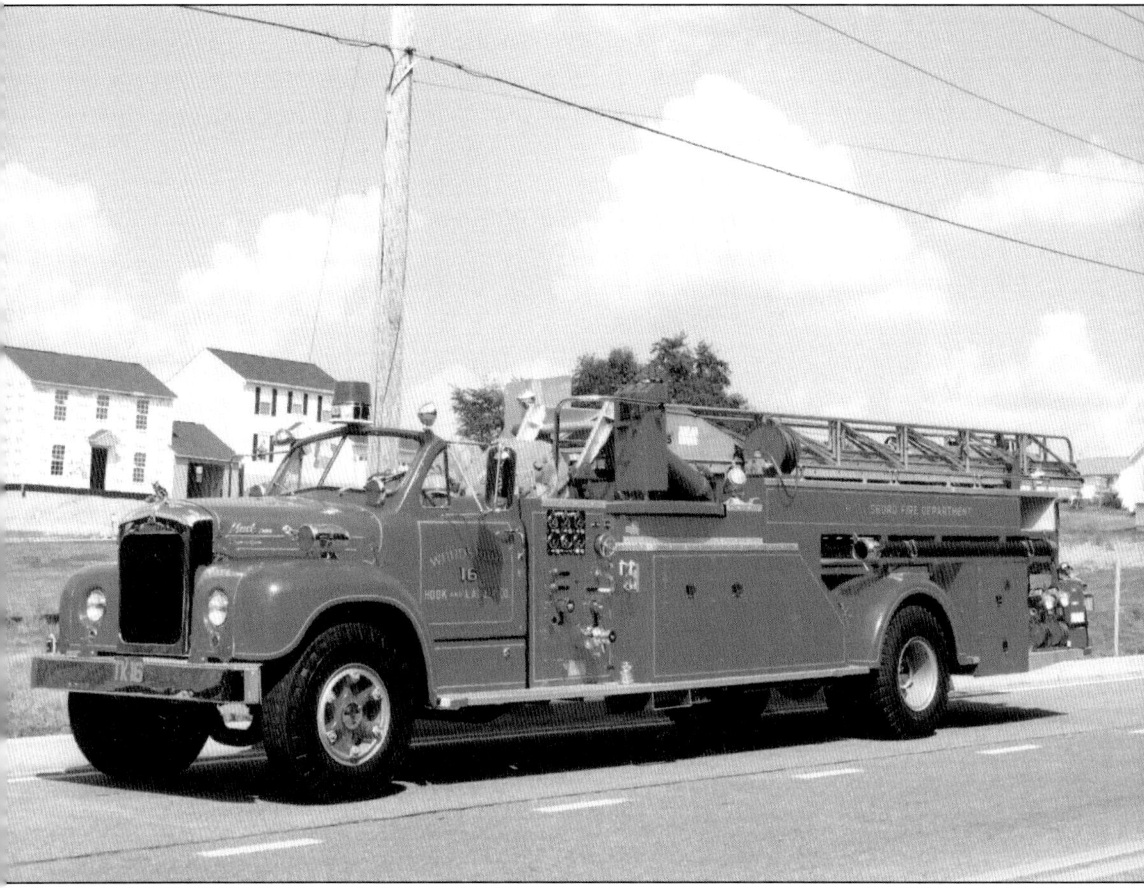

September 11, 2001, is a date in American history that will always bring visions of terror and heroism. As the events of the day unfolded, little did the members of the Woodsboro Volunteer Fire Company realize they would be called upon to play a role in the attack on the Pentagon. Truck 16 was a 1955 open-cab Mack ladder truck purchased from the Waynesboro, Pennsylvania, Fire Department. This unit was normally used on chimney fires, barn fires, and occasionally transferring to the city of Frederick. Because of the low profile and short wheel base, this ladder truck was specially called to the Pentagon to provide firefighters better access to the interior walls of the complex. The truck responded 62 miles one way to work with hundreds of firefighters from throughout the Washington area to assist in the worst terror attack in our nation's history. Pres. George Bush sent a personal letter of thanks to the members of the company for responding to the aid of their nation. (Courtesy Warren Jenkins.)

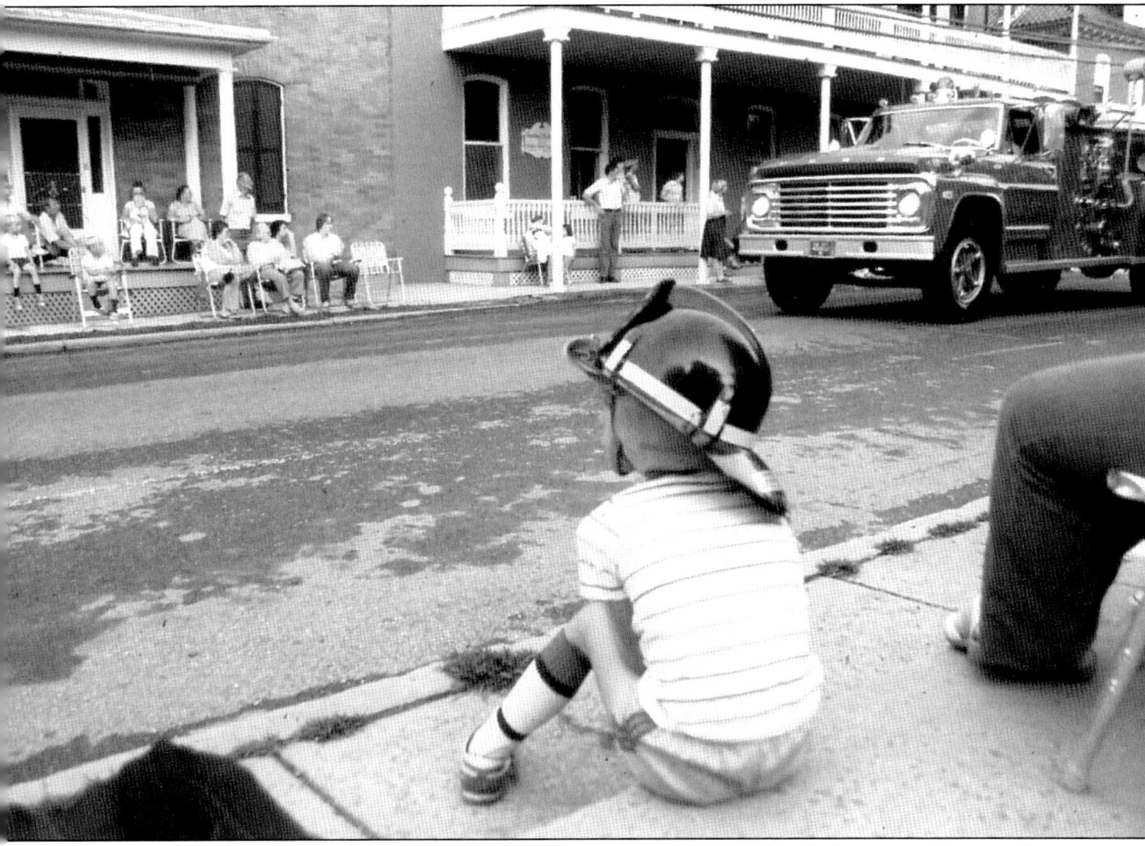

Every boy dreams of being a firefighter. Young Mike Snyder wears his helmet as a Myersville engine passes by in a parade held during the annual Woodsboro Volunteer Fire Company carnival. (Courtesy Dave Snyder.)

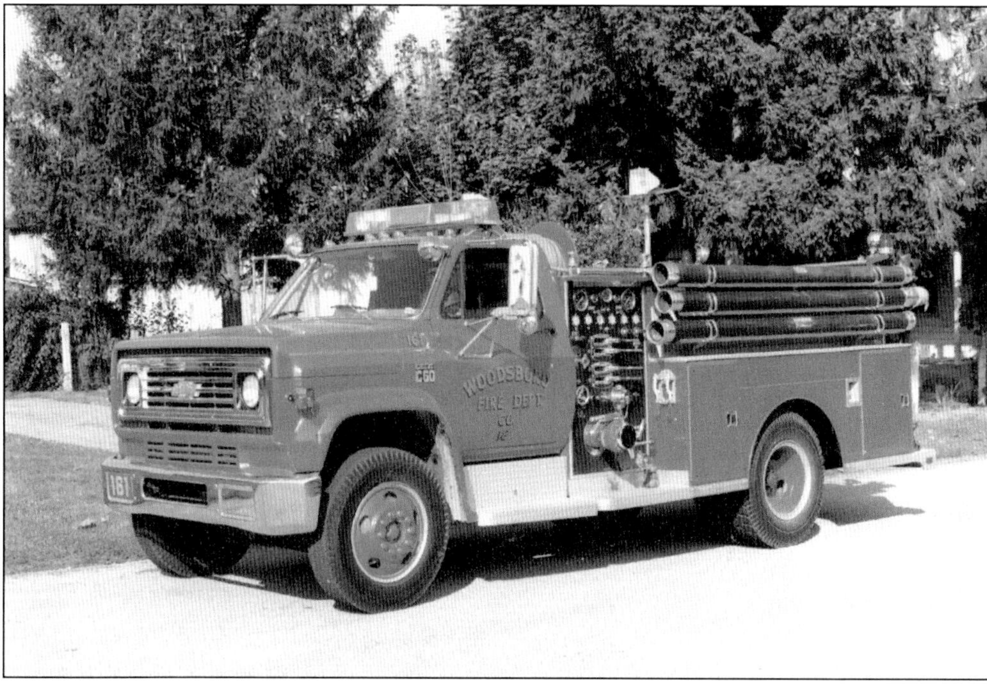

Engine 161 was a 1975 Chevrolet C60/Firemaster pumper equipped with a 750-gallon-per-minute pump and carrying 500 gallons of water. This unit served as the second due engine (the company that responds second) and primary brush fire response. (Courtesy Warren Jenkins.)

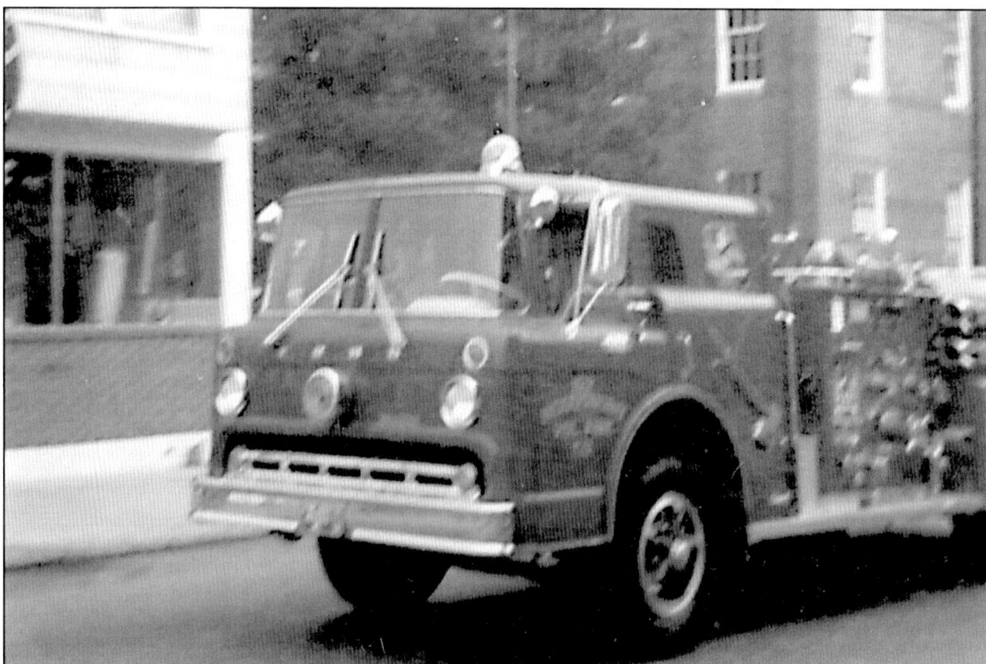

The second engine purchased by the Woodsboro Volunteer Fire Company was a 1964 Ford with an American LaFrance Body. The unit had a 750-gallon-per-minute pump and carried 500 gallons of water. (Courtesy H. Richard Hahn.)

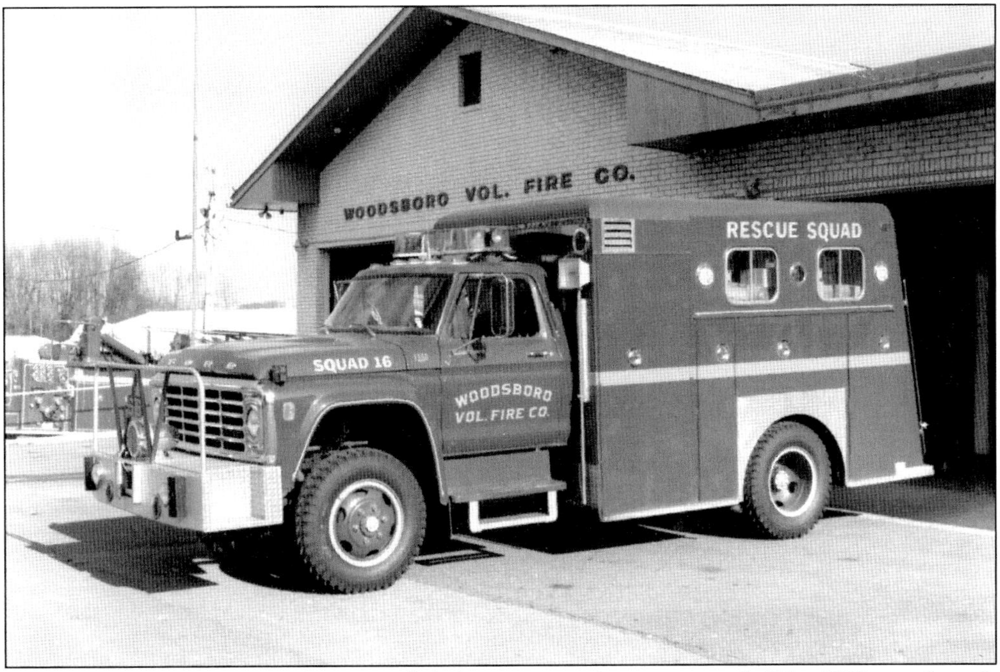

For many years, Woodsboro has provided the services of a rescue truck. The 1977 squad truck was a Morysville body on a Ford chassis and remained in service until the new squad truck was placed in service in 2002. (Courtesy Warren Jenkins.)

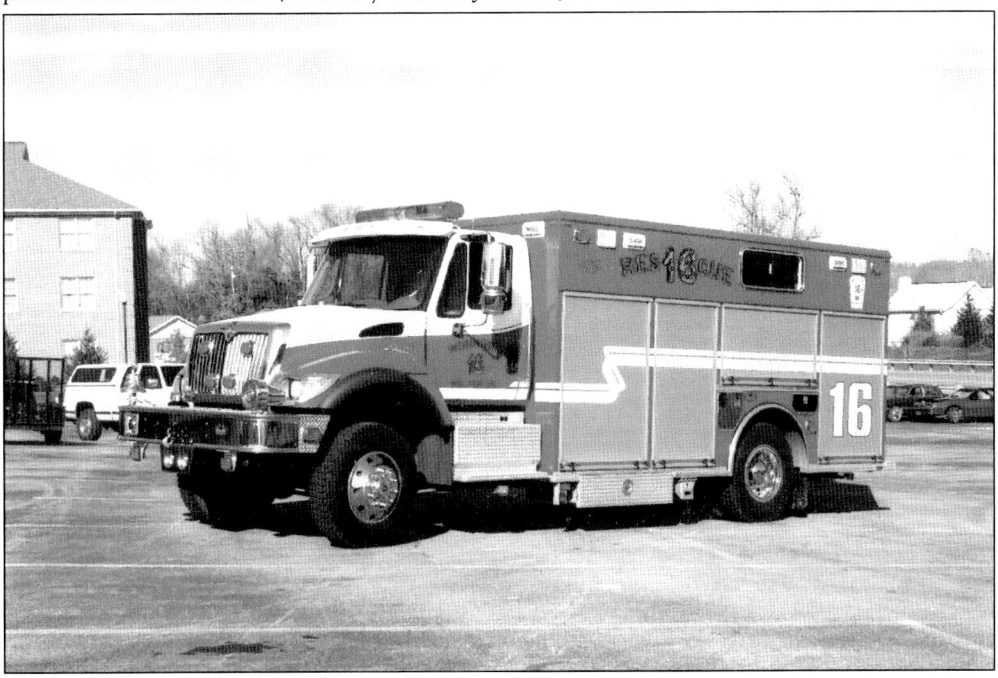

This 2002 International Kovatch is the latest addition to the Woodsboro fleet. Not only is the unit equipped for all-wheel-drive, it is also designed to serve as a basic life support ambulance. The unit has been inspected and can serve as an ambulance and even transport patients if necessary. (Courtesy Warren Jenkins.)

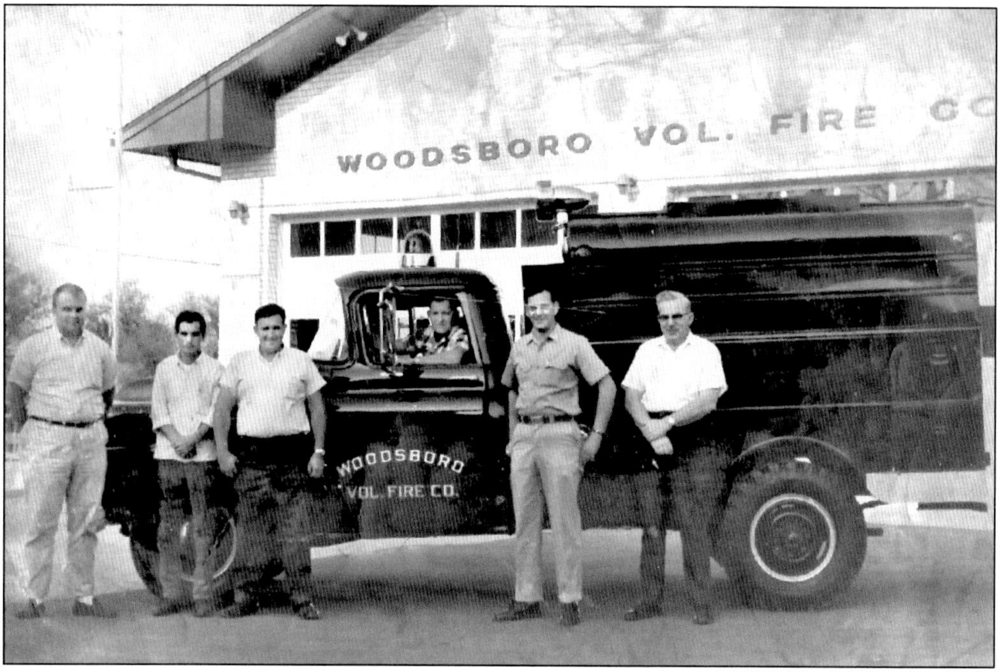

The members of Woodsboro built the first rescue truck used by the company. Here, from left to right, are Mike Staley, unidentified, Jesse Smith, Donald Gladhill, Jerry Crum, and Dorsey Rippeon with the first all-wheel-drive rescue truck at the Woodsboro Fire Company quarters. (Courtesy Woodsboro Volunteer Fire Company [WBFC].)

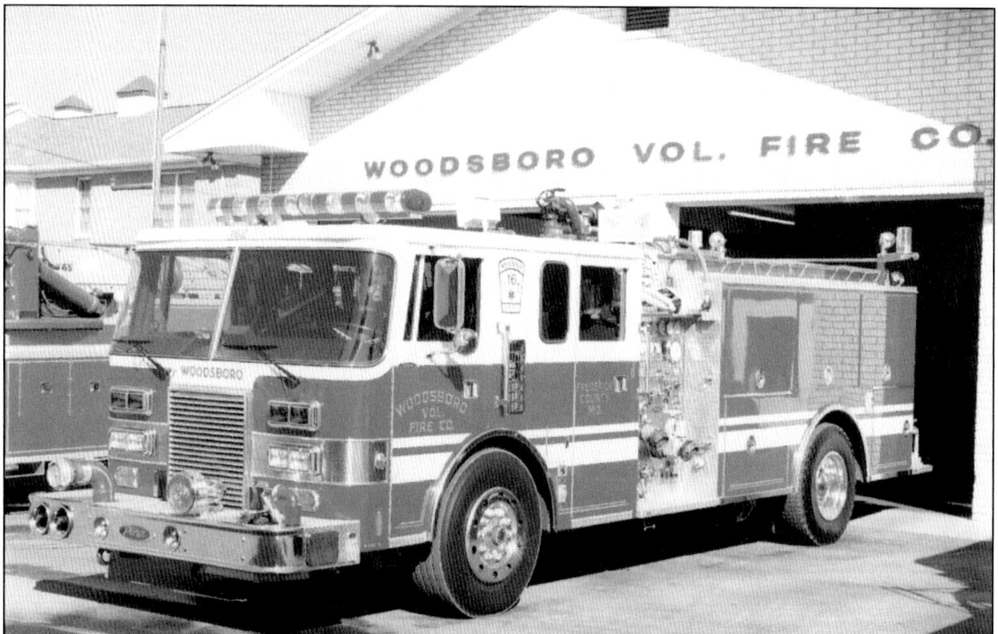

In 1989, the Woodsboro Volunteer Fire Company purchased this Pierce Lance engine with a 1,250-gallon-per-minute pump and 1,000 gallons of water. This engine replaced the 1964 Ford American LaFrance engine. (Courtesy Warren Jenkins.)

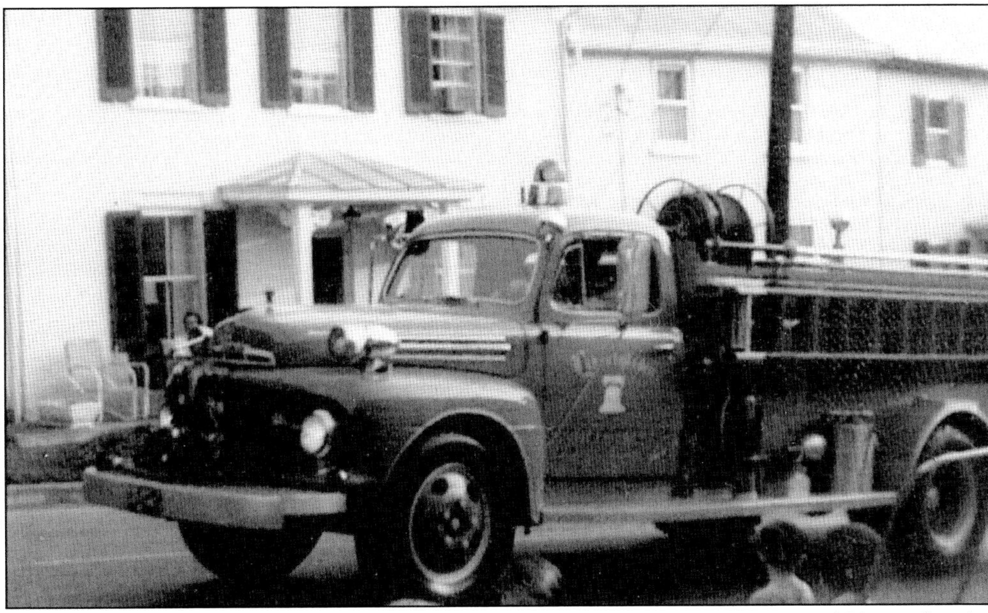

Though the United Volunteer Fire Company of Libertytown was in existence in 1890, another fire company was not organized in Libertytown until 1964. The first engine purchase by Libertytown Volunteer Fire Department was this 1952 Ford engine purchased from the Ridge Volunteer Fire Company in St. Mary's County. The new company did not have to wait long, as the first alarm was for a plane crash at the Burrier farm on Clemsonville Road during a severe thunderstorm. (Courtesy Libertytown Volunteer Fire Department [LVFD].)

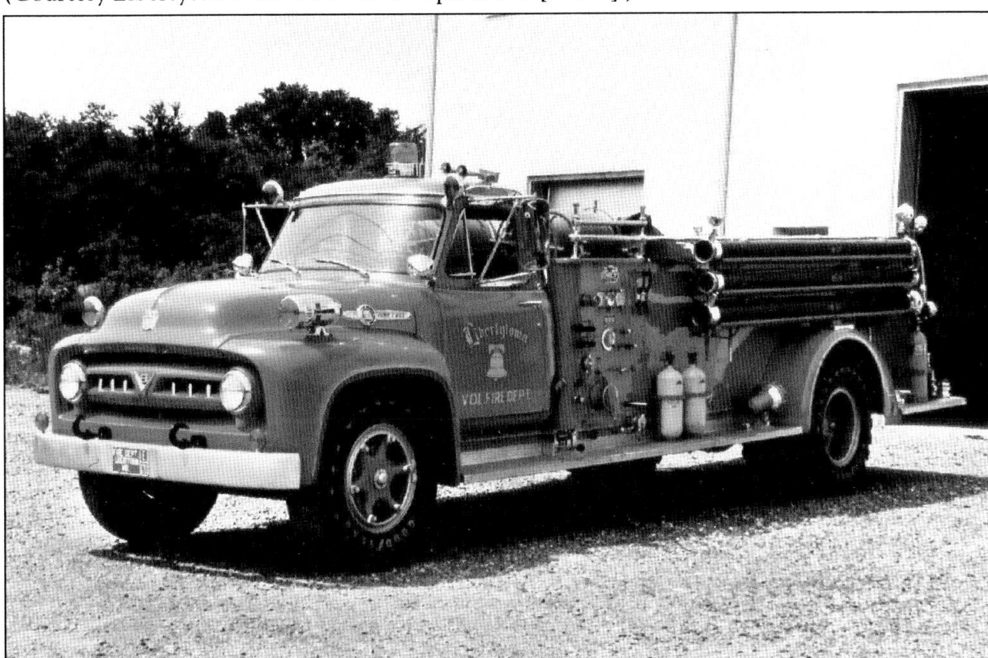

The new company found a second engine to help protect the community. This 1953 Ford had a mid-ship pump rather than the front-mounted pump on the first engine purchased. (Courtesy LVFD.)

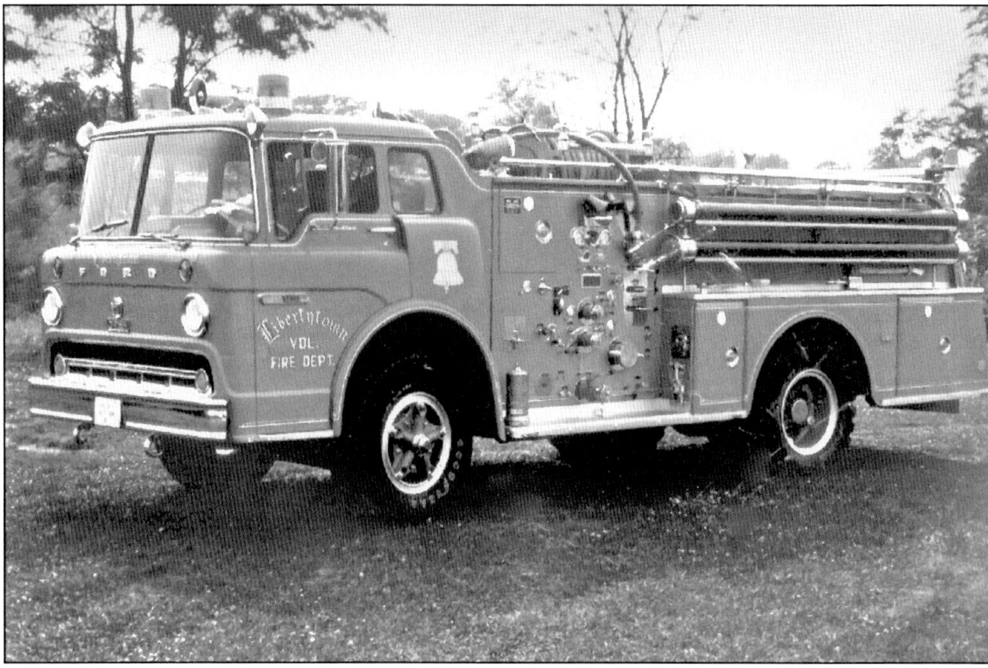

Engine 173 was a 1966 Ford Oren purchased used; it had been involved in an accident and rebuilt/retitled as a 1970 engine. This unit had a 750-gallon-per-minute pump and carried 500 gallons of water. The unit remained in service until the mid-1980s. (Courtesy LVFD.)

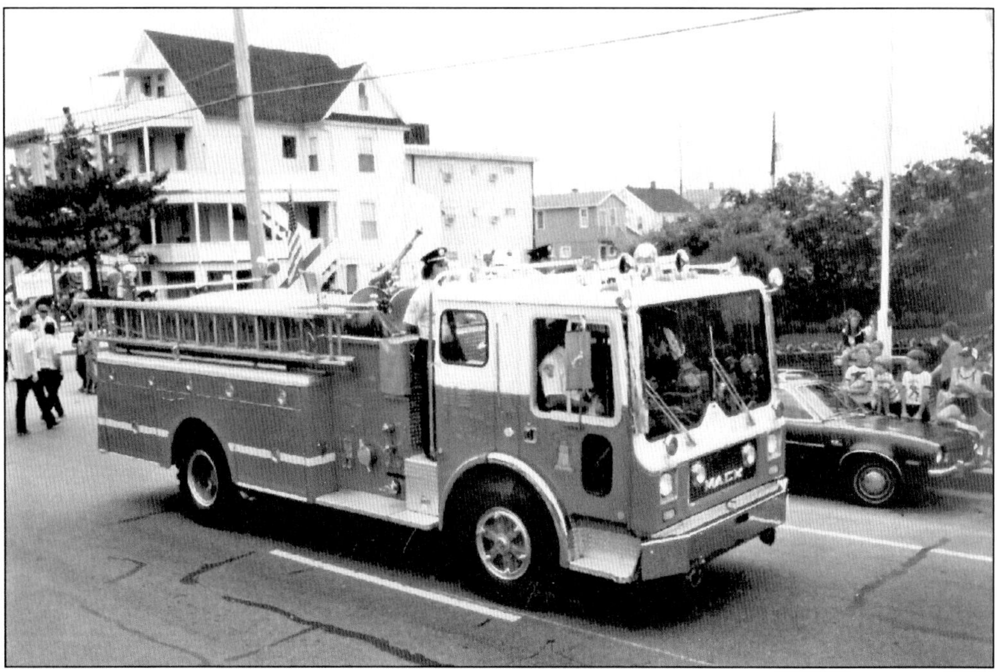

In 1980, the Libertytown Volunteer Fire Department purchased Engine 171, this MC model Mack. This unit was the first MC Mack pumper in Maryland. This picture was taken at the parade of the Maryland State Firemen's Association in Ocean City. The engine won first place. (Courtesy LVFD.)

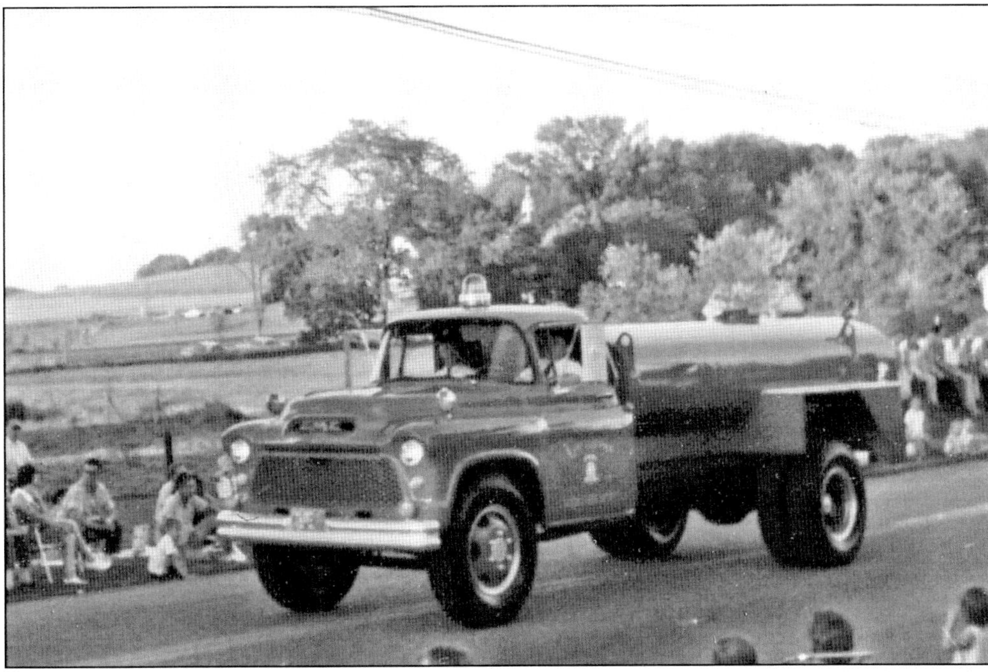

After two major fires in the 1960s, it was evident water supply would be critical for the new fire company. The first tanker purchased was this 1957 GMC Tanker. This vehicle was later sold to the Graceham Volunteer Fire Company. (Courtesy LVFD.)

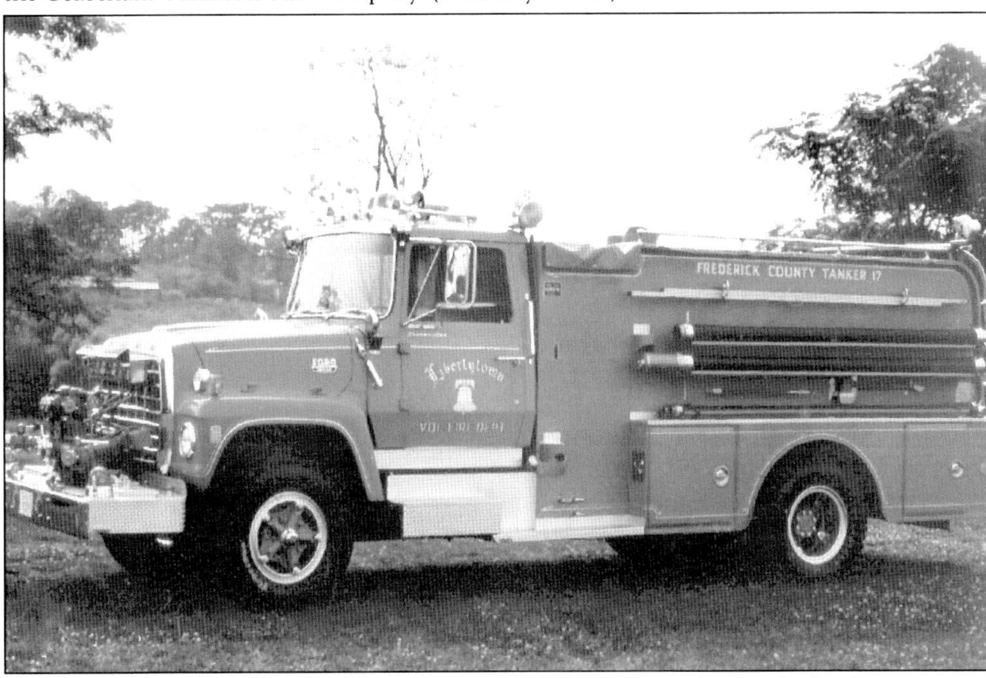

This Ford Louisville tanker replaced the GMC tanker in 1973. The unit had a front-mounted pump and carried a full compliment of hose for firefighting and could respond as an engine if necessary. This tanker was the last unit built by Oren to be placed in service in Frederick County. (Courtesy LVFD.)

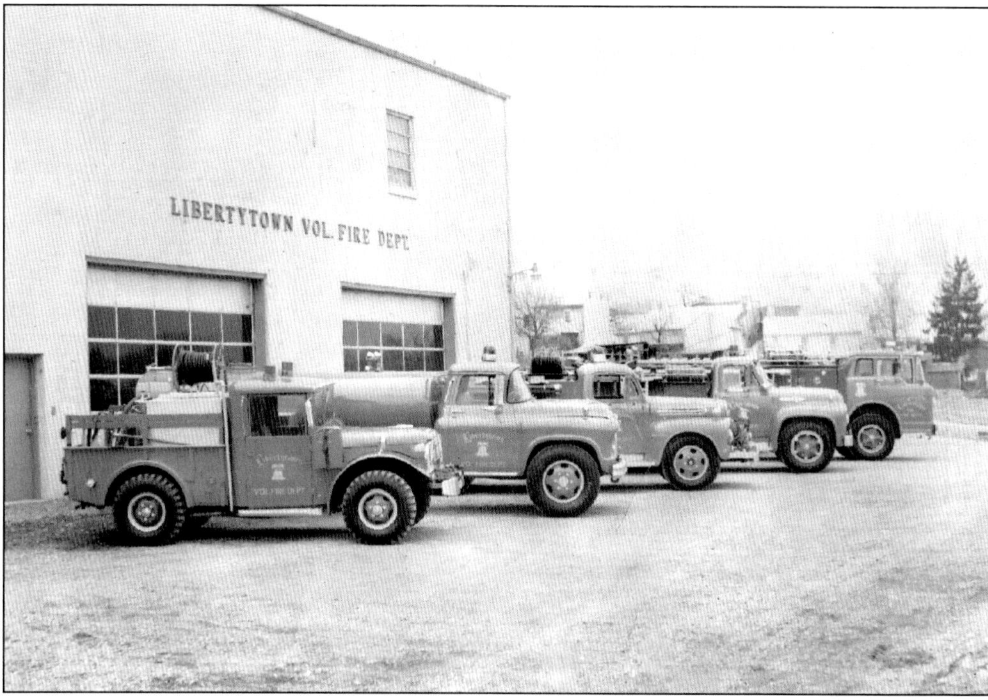

These pictures show the Libertytown equipment today and yesterday. The picture at the top shows the equipment in service in 1970, while the picture below shows the equipment in 2005. The fire company expanded to include ambulance service in 1979 and today responds to over 700 total calls a year. (Courtesy LVFD.)

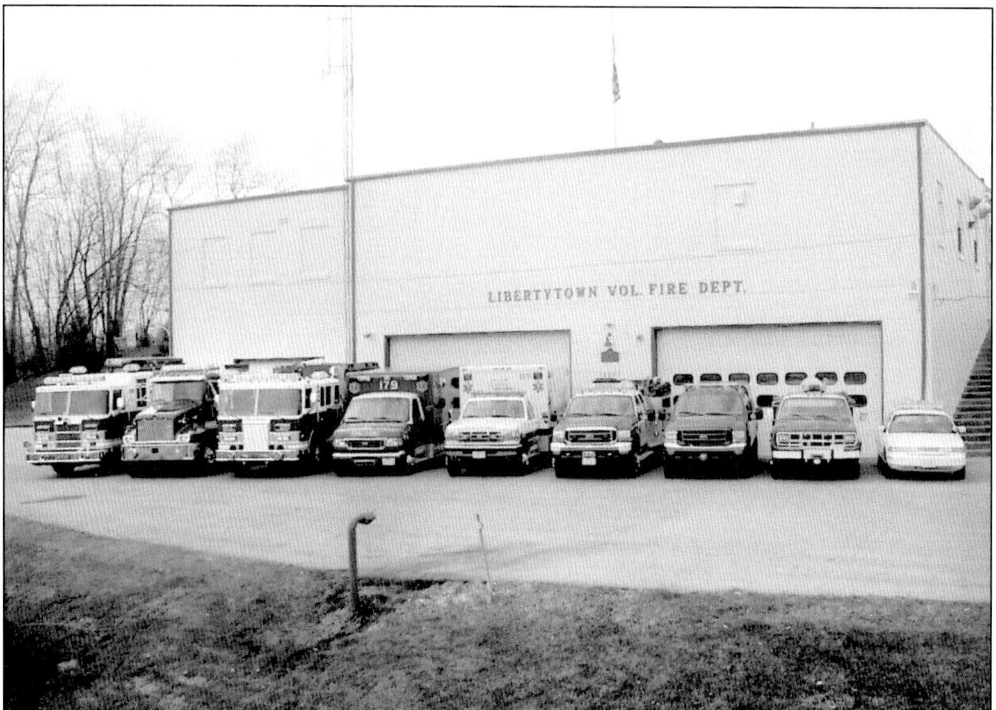

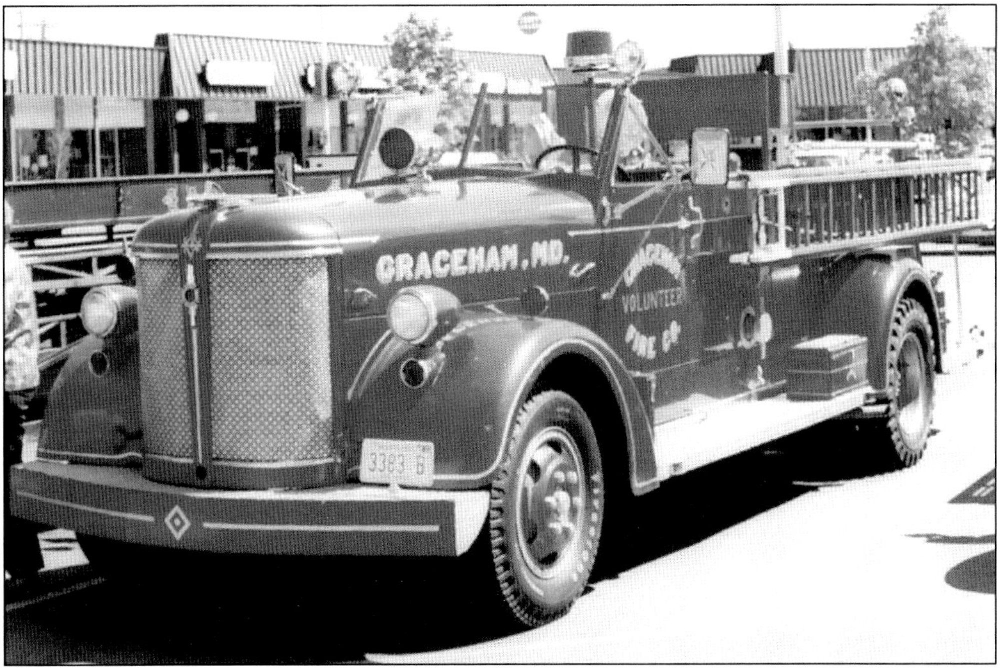

There has been fire protection in Graceham since at least 1939. In fact, there were actually two companies at different times. The modern Graceham fire company was organized in April 1960. The 1947 American LaFrance shown here was the first engine owned by the company. It was purchased from the Branchville Volunteer Fire Department in Prince George's County. It was originally a "plain Jane" engine built for the U.S. military. (Courtesy Warren Jenkins.)

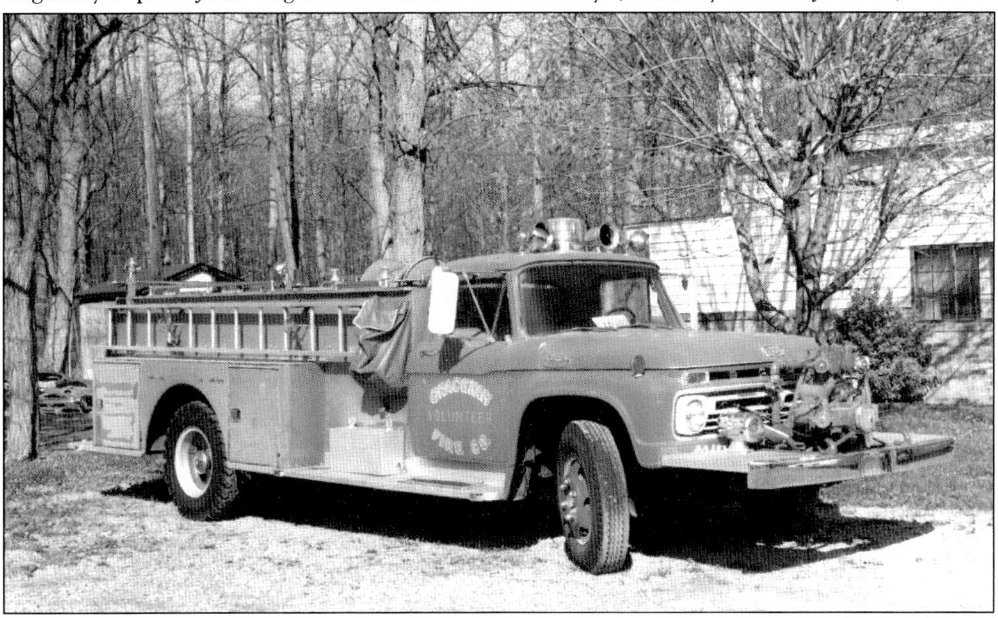

This 1966 Ford Darley was the first new piece of apparatus purchased by the Graceham Volunteer Fire Company. This engine was relatively inexpensive; however, it was very versatile. The unit had a 750-gallon-per-minute front-mounted pump and carried 500 gallons of water. (Courtesy Warren Jenkins.)

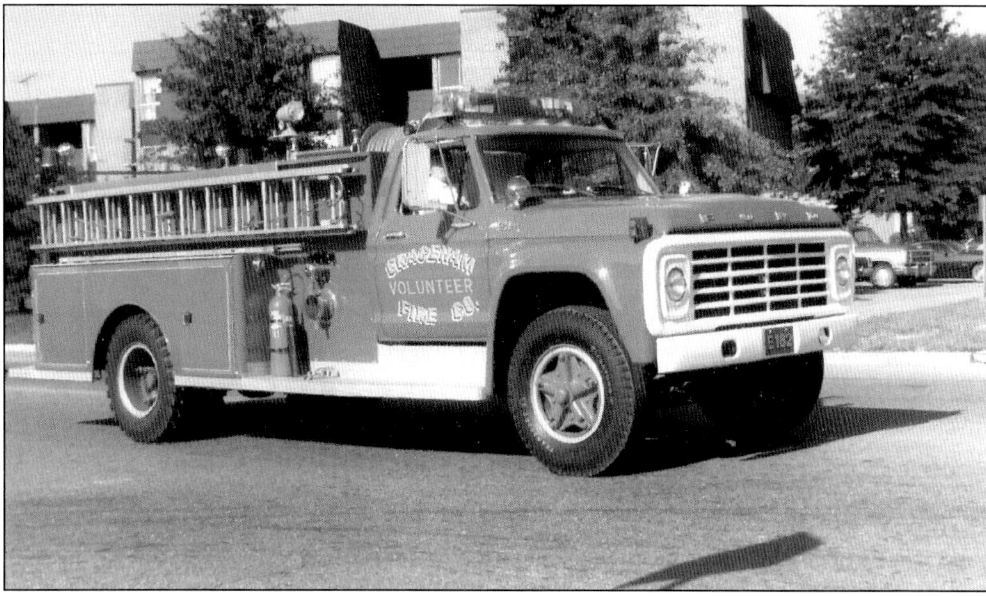

Engine 182 was the second new Darley to be added to the Graceham fleet. This 1980 Ford F800 had a 1,000-gallon-per-minute pump and 750 gallons of water. (Courtesy Graceham Volunteer Fire Company [GVFC].)

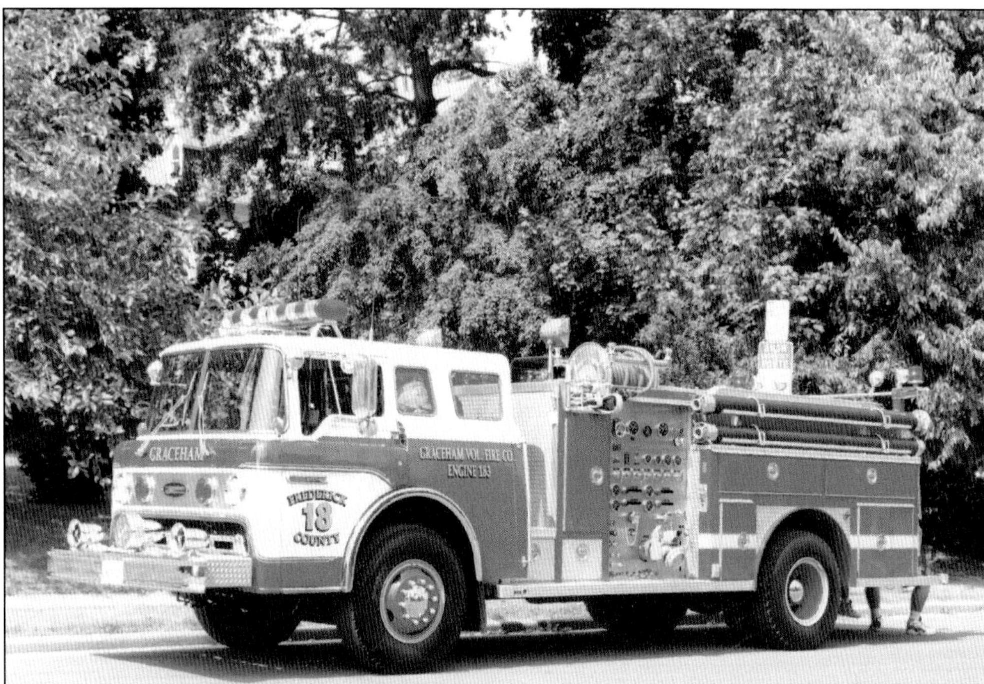

The Graceham Volunteer had two similar Fords in service at the same time; shown is Engine 183, a 1980 Ford Emergency One pumper purchased from the Urbana Volunteer Fire and Rescue Company. The engine has a 1,000-gallon-per-minute pump and carries 750 gallons of water. (Courtesy Warren Jenkins.)

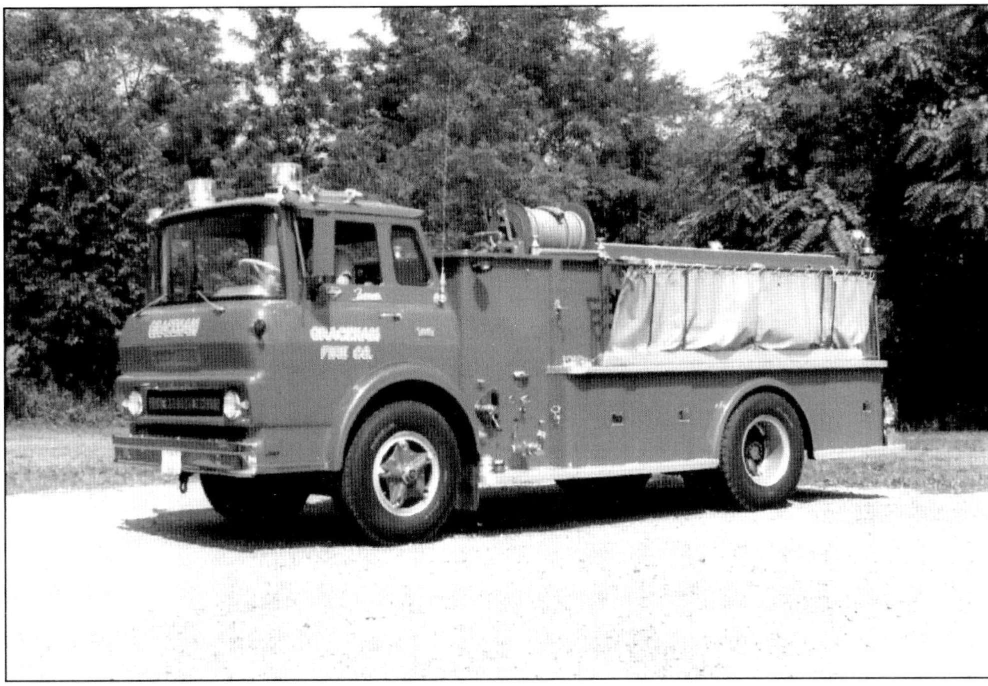

Graceham has operated a tanker for over 30 years. The first tanker was a GMC purchased from Libertytown and had a variety of chassis until replaced. It was eventually replaced by this tanker purchased from Mansfield, Connecticut. Tanker 18 was a 1966 GMC Farrar with a 375-gallon-per-minute pump and a tank that carried 1,600 gallons of water. (Courtesy GVFC.)

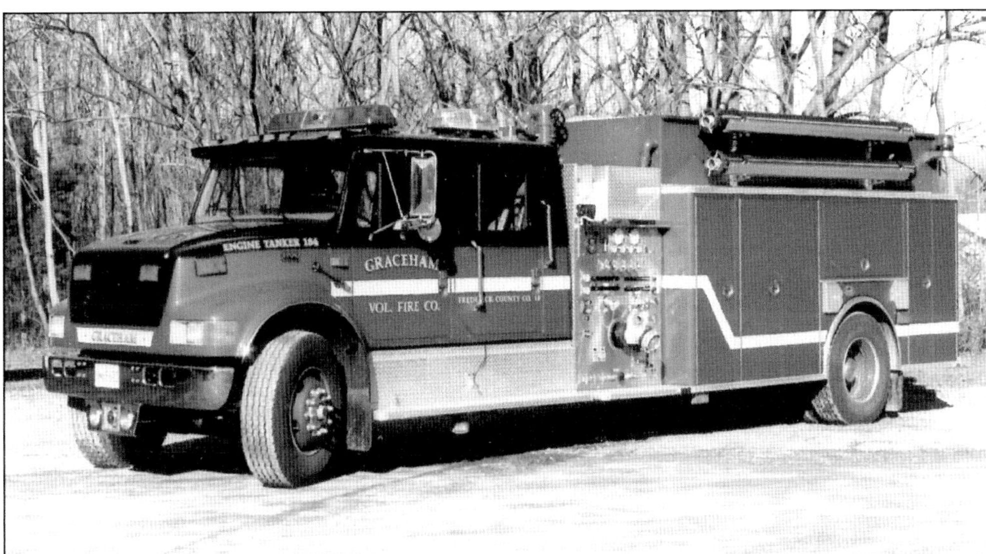

The new Engine Tanker 184 is a 2004 International pumper/tanker with a fully enclosed cab and carrying 1,800 gallons of water. With this engine, Graceham also implemented a new color scheme of black over a dark maroon red on the newly purchased equipment. (Courtesy GVFC.)

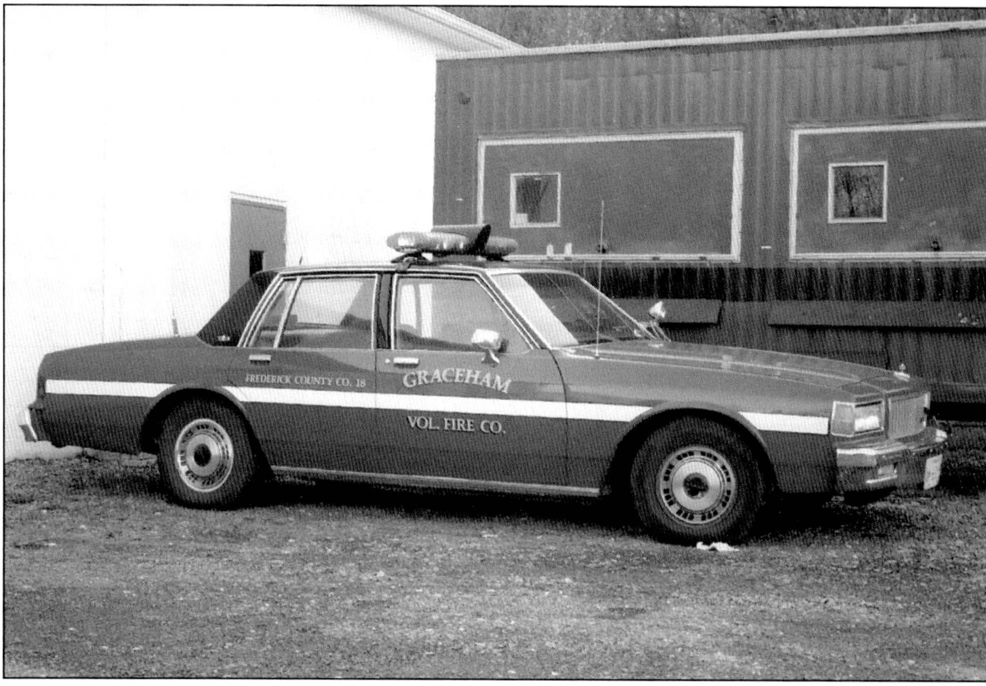

The distinctive black and red color was also added to the chief/duty vehicle of the fire company. This vehicle has recently been replaced by a vehicle donated by the Mount Airy Volunteer Fire Company in Carroll County. (Courtesy Warren Jenkins.)

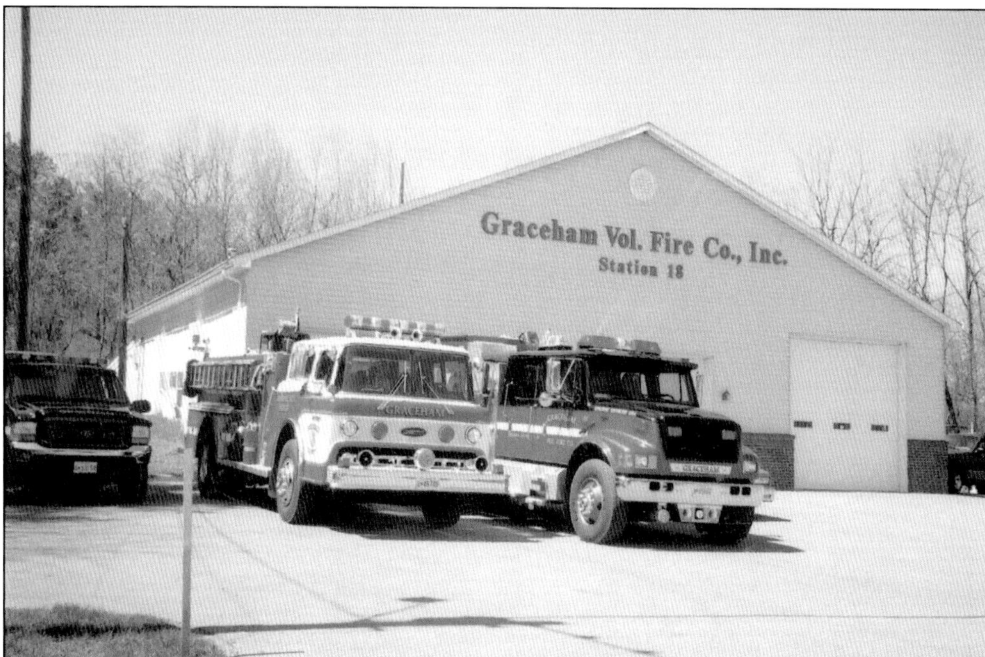

From left to right, Special Unit 18, Engine 183, and Engine Tanker 184 are shown in front of the Graceham Volunteer Fire Company building after the building renovation. (Courtesy Warren Jenkins.)

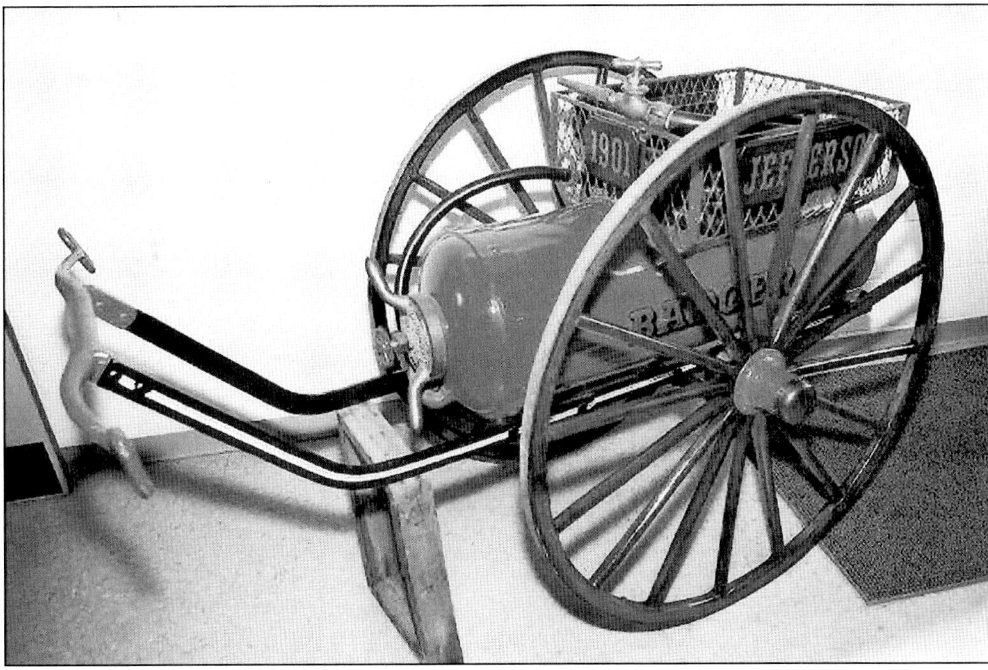

In 1901, two Badger pressurized hand-pulled extinguishers were purchased to protect the town of Jefferson. One of the extinguishers has been restored and is still pulled in parades and displayed at local events. The first fire engine was actually owned by the Braddock Heights Volunteer Fire Company with the beginning of the Jefferson substation. (Courtesy Dave Snyder.)

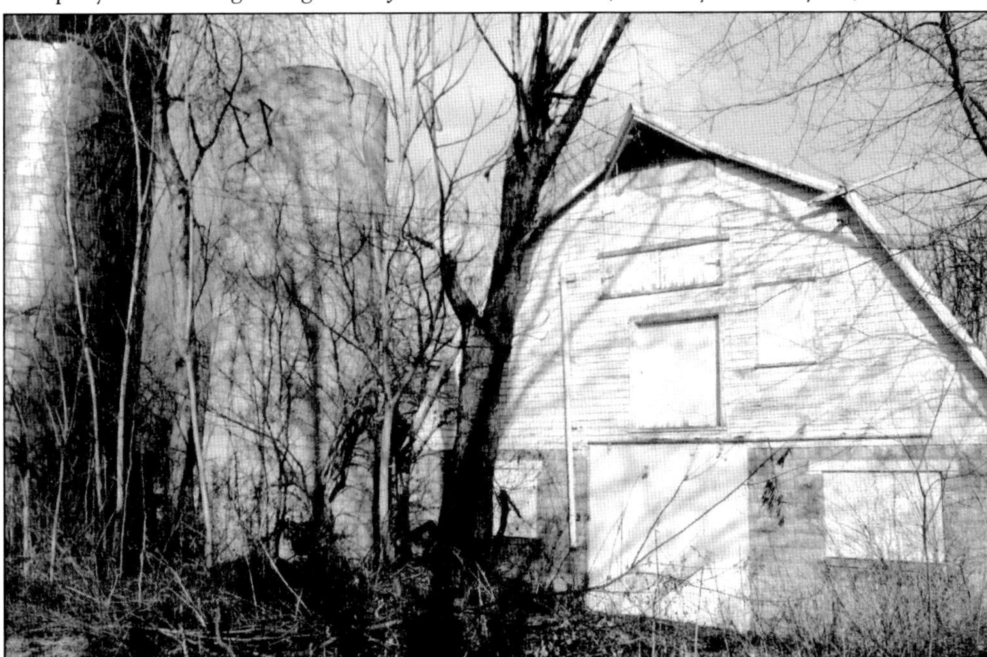

The original site of the first fire station was this barn, located on the west end of town. The fire station was also located at the former post office site before being moved to the present location on the Ruritan grounds. (Courtesy Dave Snyder.)

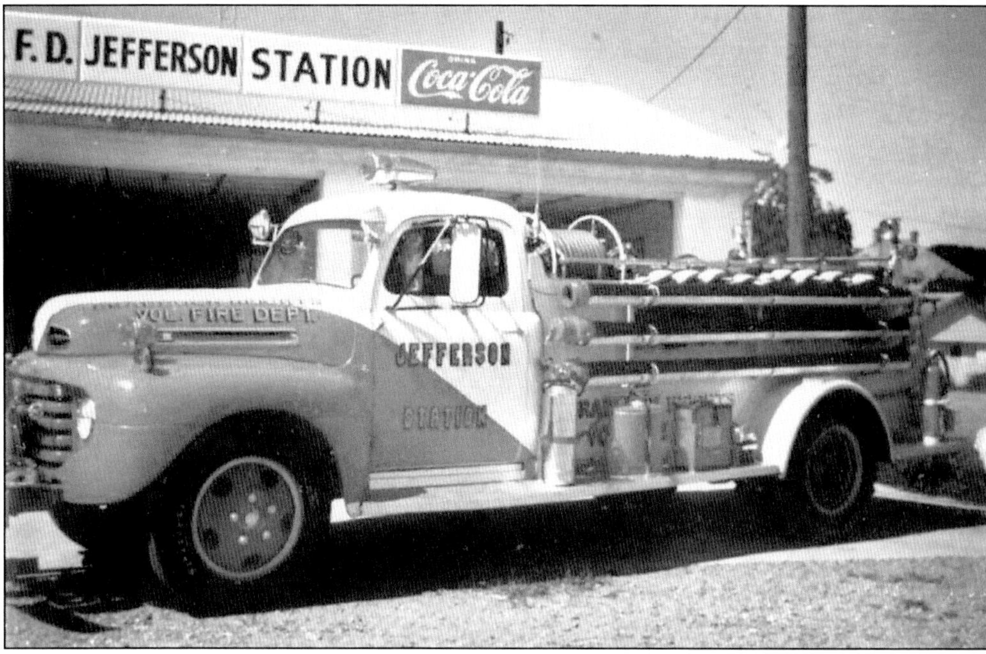

The first pumper owned by the Braddock Heights Volunteer Fire Company is shown in front of the present station location before the Jefferson Volunteer Fire Company was officially organized. Note the BHVFC on the sign and Jefferson Station on the doors of the engine. (Courtesy Dave Snyder.)

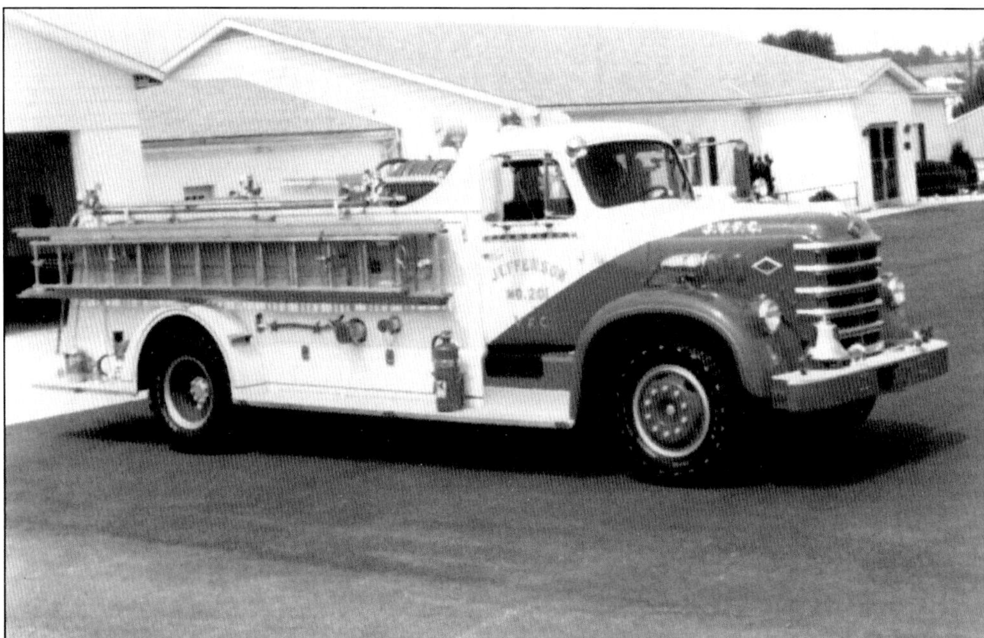

This 1957 Diamond-T engine was purchased to be used at the Jefferson station. It was needed to replace the one engine destroyed in a fire at the fire station in 1968. Though the station received smoke damage, the engine was nearly destroyed. It was believed an electrical fire in the fire engine was the cause of the blaze. (Courtesy Dave Snyder.)

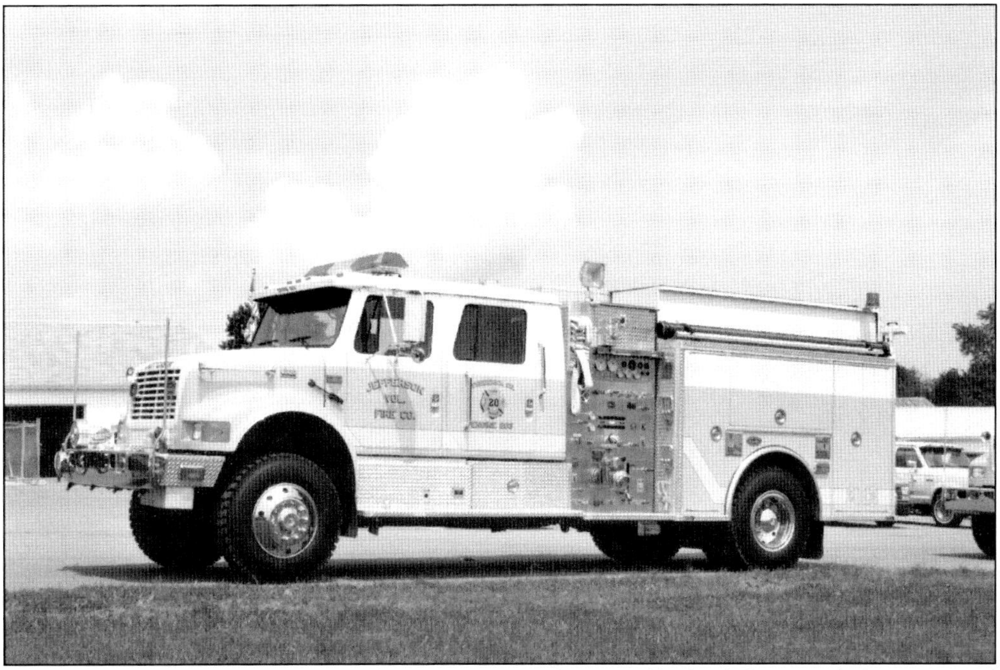

Engine 203 is an International multi-purpose unit having all-wheel-drive and also equipped as a Class A full-service pumper. This unit also has a fully enclosed four-door cab. (Courtesy Warren Jenkins.)

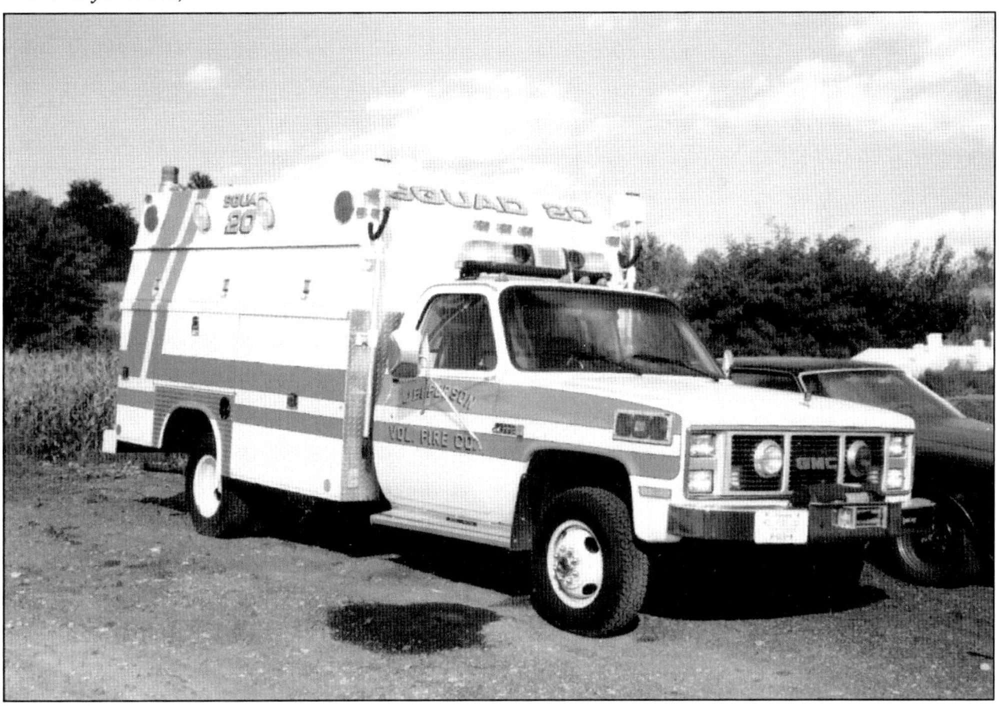

Protecting many miles of U.S. Route 340, the Jefferson Volunteer Fire Company placed in service a 1988 GMC Reading Body Light Duty Squad. The fire company has since updated the rescue service to include an air cascade system on the present rescue unit. (Courtesy Warren Jenkins.)

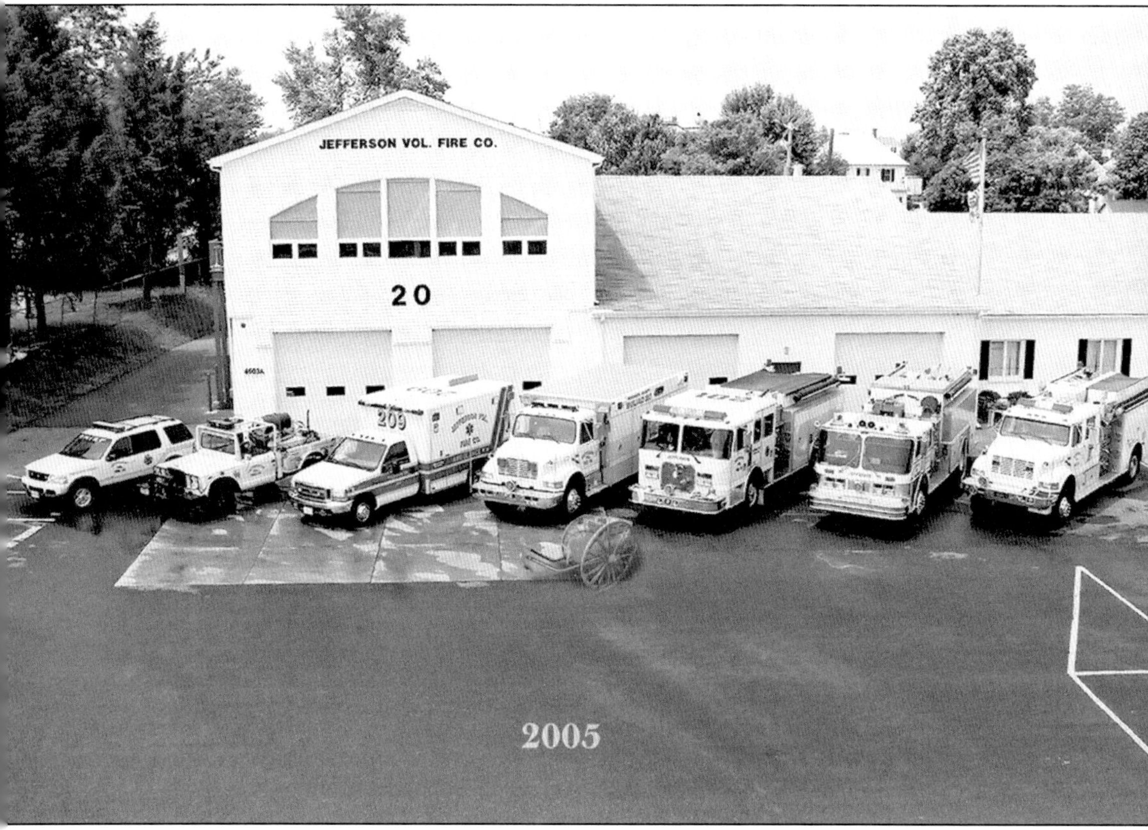

The Jefferson Volunteer Fire Company recently completed a new building that was the first fire station in the United States to utilize a fog high-mist sprinkler system. The company now includes both ambulance and rescue service. The firehouse property is actually owned by the Jefferson Ruritan Club, an organization that has supported the fire company from the beginning. (Courtesy Dave Snyder.)

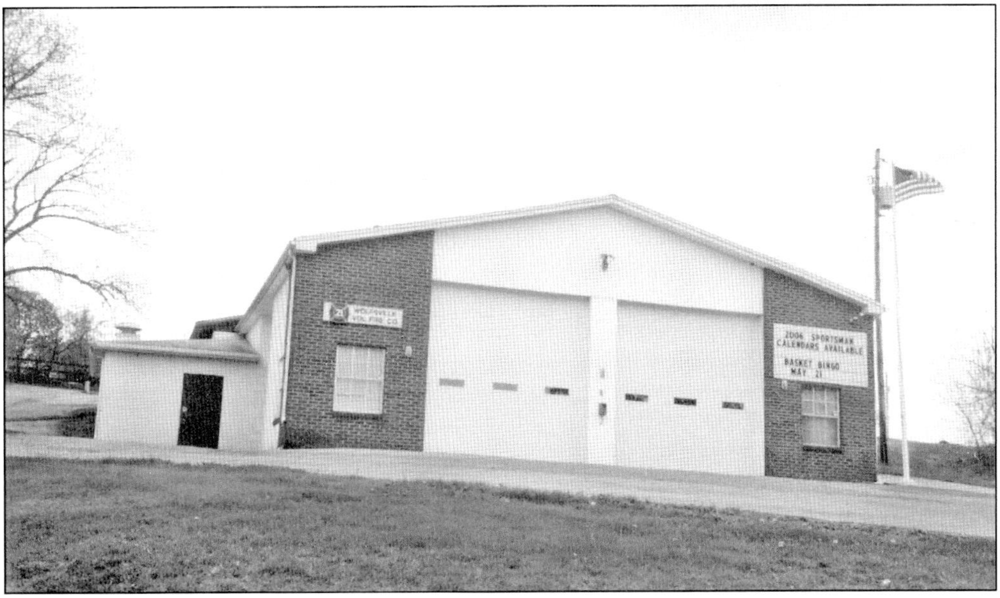

In 1968, the Myersville Volunteer Fire Company helped to organize a fire station in the town of Wolfsville. The Wolfsville Holding Company was incorporated to serve as the administrative and financial arm of the Myersville Volunteer Fire Company substation. The firehouse was built for a cost of $10,033. In December 1979, the Wolfsville Volunteer Fire Company was organized and separated from the Myersville Company. The holding company was dissolved. (Courtesy Warren Jenkins.)

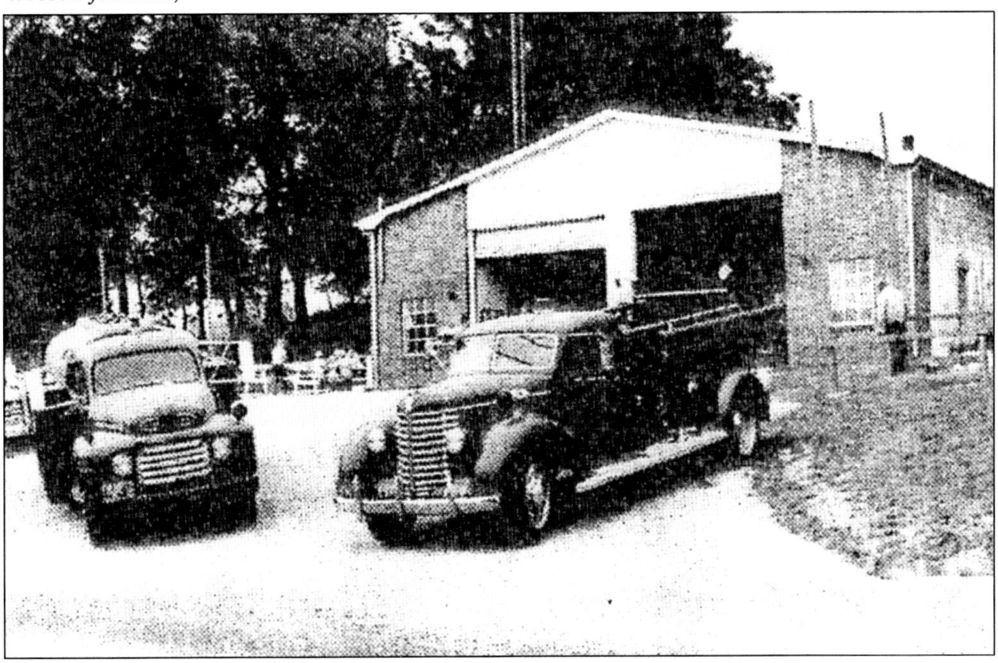

The first equipment housed at the Wolfsville substation was the Myersville 1941 Diamond-T engine. The tanker shown was a 1949 Ford two-ton chassis with a 1,200-gallon tank purchased from Rocky Ridge and placed on the truck. (Courtesy Wolfsville Volunteer Fire Department [WFC].)

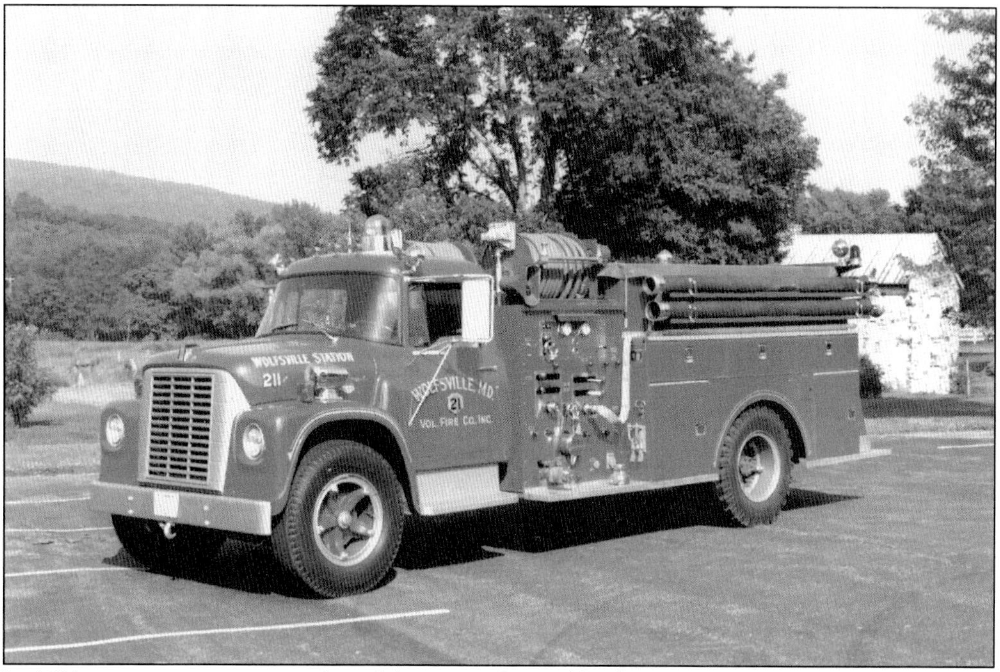

In 1971, a new engine was purchased for use at the Wolfsville station. A 1971 International was purchased for a total cost of $22,888. The unit was a Maxim body with a 750-gallon-per-minute pump, and it carried 500 gallons of water. (Courtesy Warren Jenkins.)

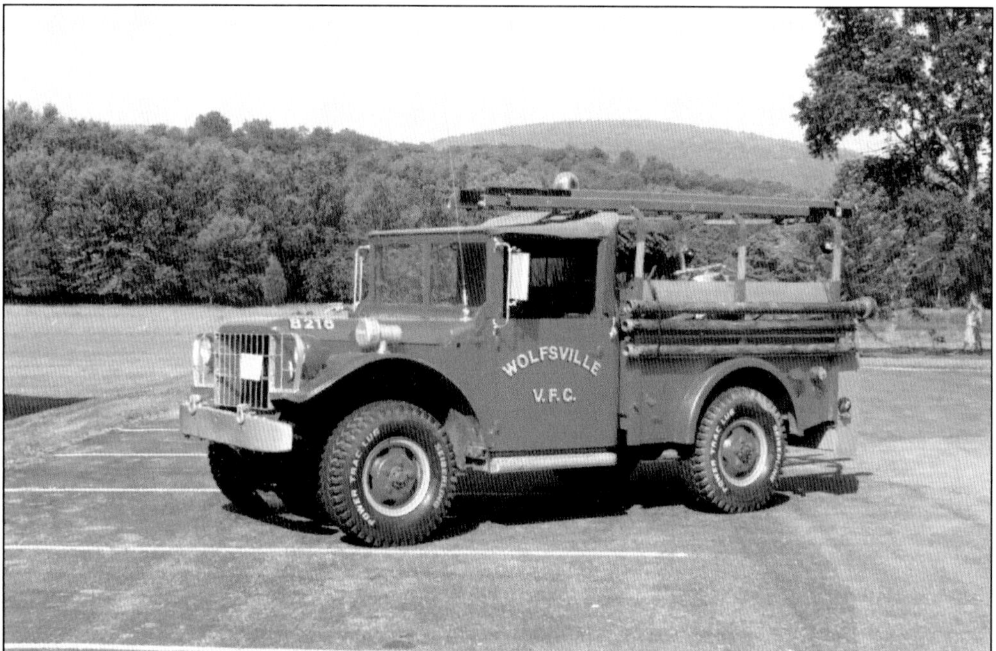

Located high in the Catoctin Mountains, Wolfsville had a need to purchase a rugged brush unit as soon as the company was in service. One of the first Wolfsville brush units was this mighty 1951 four-wheel-drive Dodge Powerwagon. The unit had a small pump and carried 60 gallons of water. (Courtesy Warren Jenkins.)

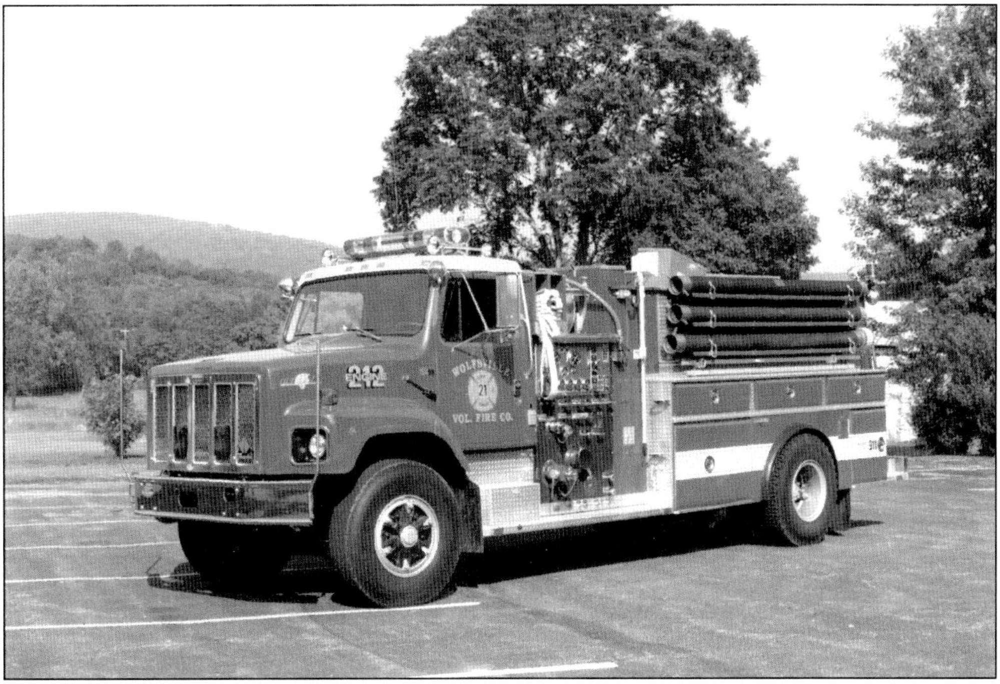

Engine/Tanker 212 was a 1987 International Grumman engine tanker. The unit had a 1,000-gallon-per-minute pump and carried 1,500 gallons of water. This unit was recently replaced by a 2004 Grumman engine tanker. (Courtesy Warren Jenkins.)

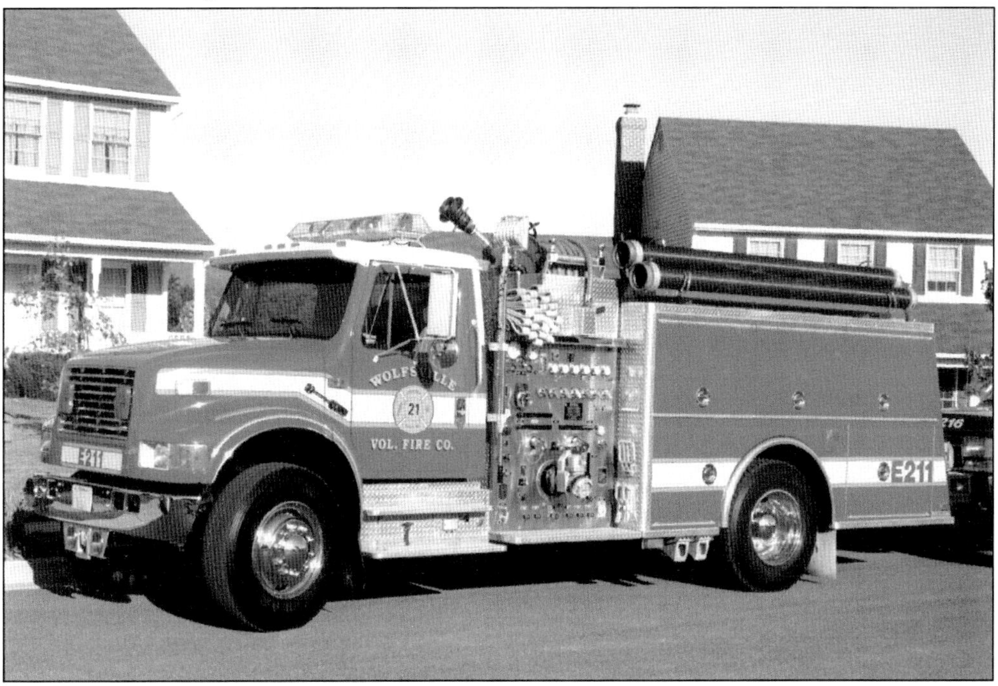

Engine 211 is a 1992 International KME with a 1,250-gallon-per-minute pump that carries 750 gallons of water. The engine was also equipped as a light rescue unit with extrication equipment for auto accidents. (Courtesy Warren Jenkins.)

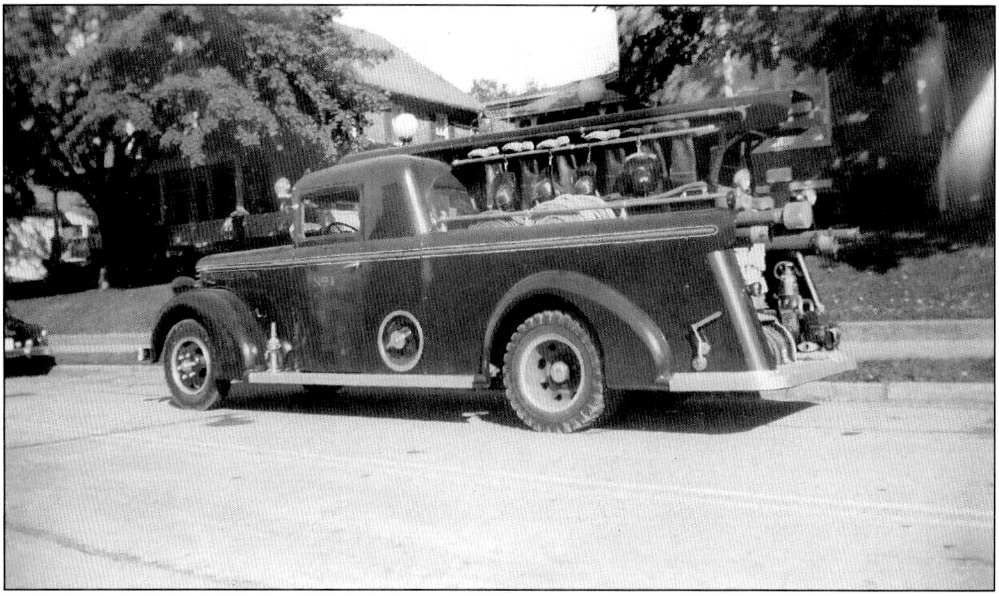

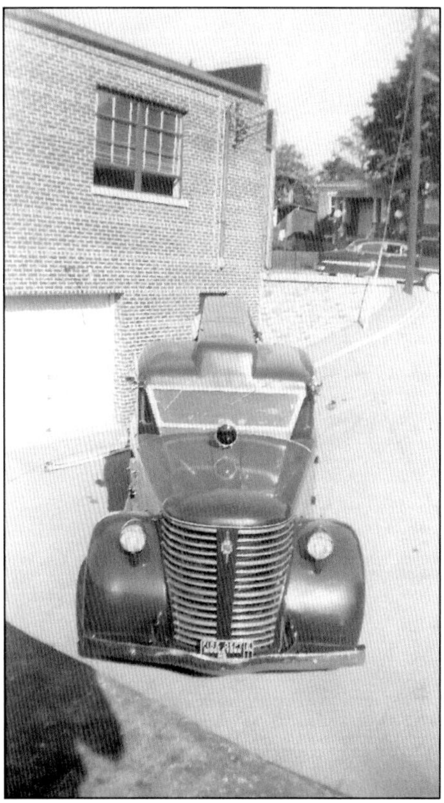

The sleek design of the 1939 American LaFrance is shown when it was delivered new to the Middletown Volunteer Fire Company. The engine was especially designed for optimum storage of equipment by enclosing a ladder rack along the top of the engine. The unit was originally Engine 1 in Middletown and ran as the original Engine 221 for the Lewistown District Volunteer Fire Department. Both the front and side views show the aerodynamic design of the engine. Lewistown purchased the engine for $450. (Courtesy MVFC.)

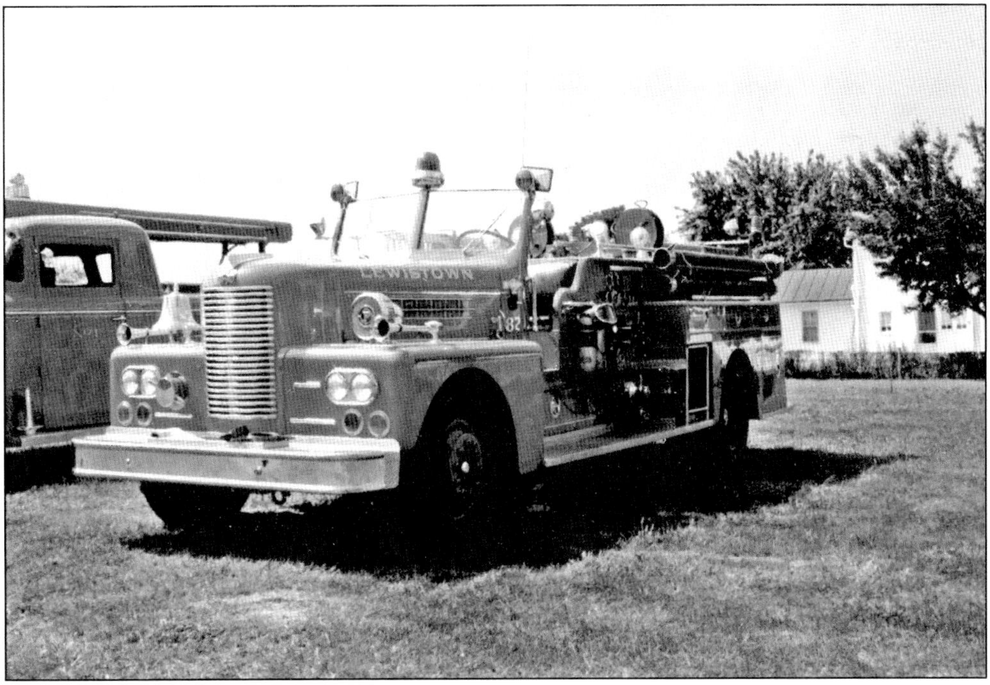

This open-cab 1961 Pirsch had a 750-gallon-per-minute pump and carried 500 gallons of water. This unit was purchased used from the Seat Pleasant Fire Department in Prince George's County. (Lewistown District Volunteer Fire Department.)

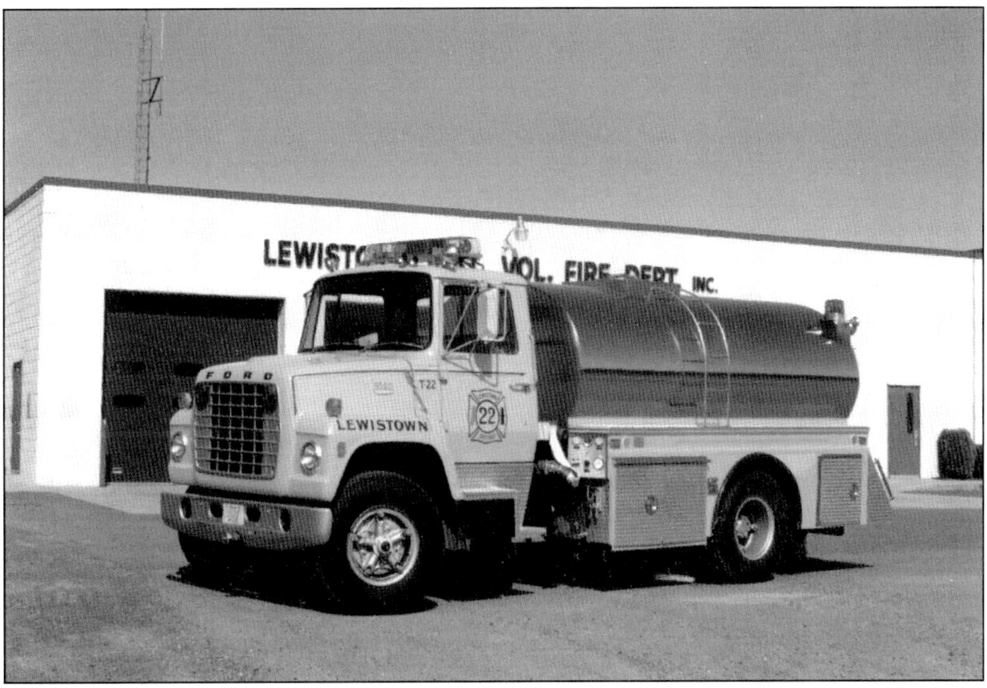

The Lewistown District Fire Department purchased this tanker new in 1978. Tanker 22 was a 1978 L800 Ford built by Four Guys in Myersdale, Pennsylvania. The tanker carried 2,000 gallons of water and had a 400-gallon-per-minute pump. (Courtesy Warren Jenkins.)

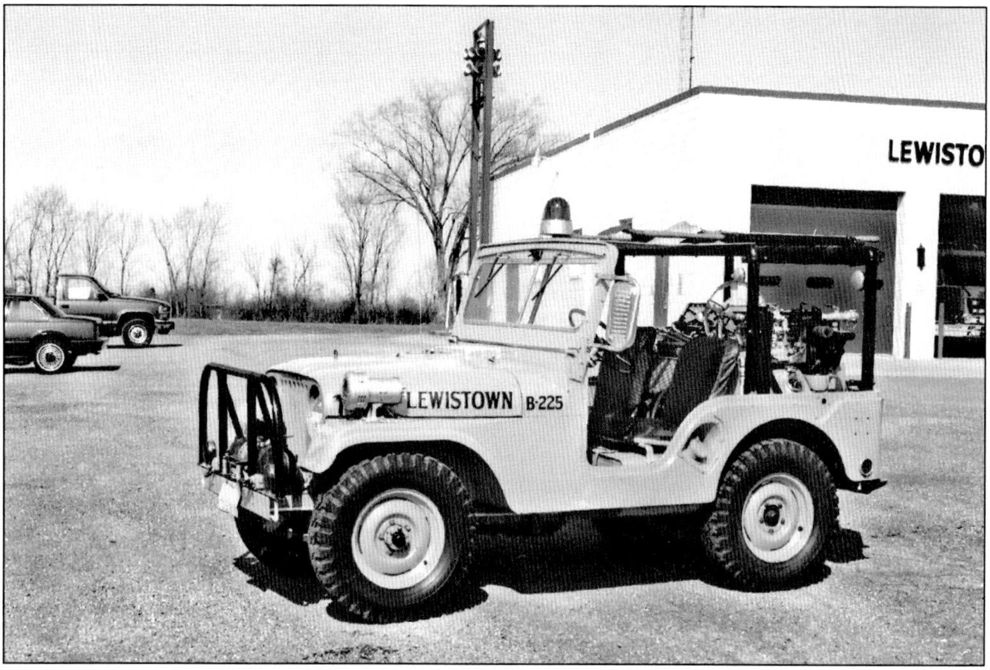

With a large area of forests, including the Frederick Watershed, in the first-response area, maneuverable brush units are a must. This 1953 CJ-5 Jeep had a portable pump and carried 75 gallons of water. (Courtesy Warren Jenkins.)

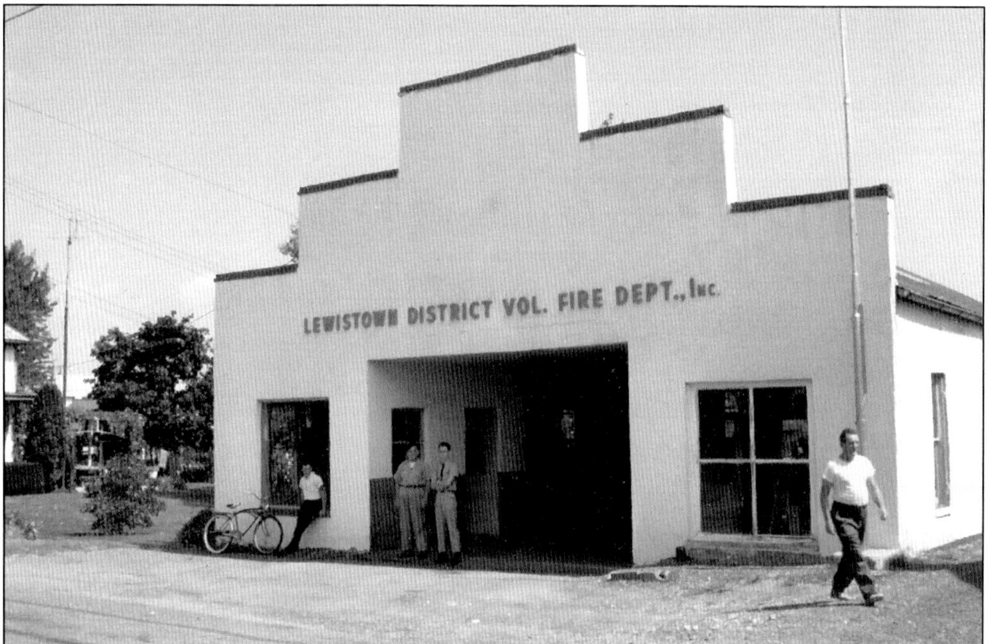

The first location for the Lewistown District Volunteer Fire Department was located on Hessong Bridge Road north of the present location. The garage was originally owned by Jim Grimes, a charter member and past president of the fire department. (Courtesy Lewistown District Volunteer Fire Department.)

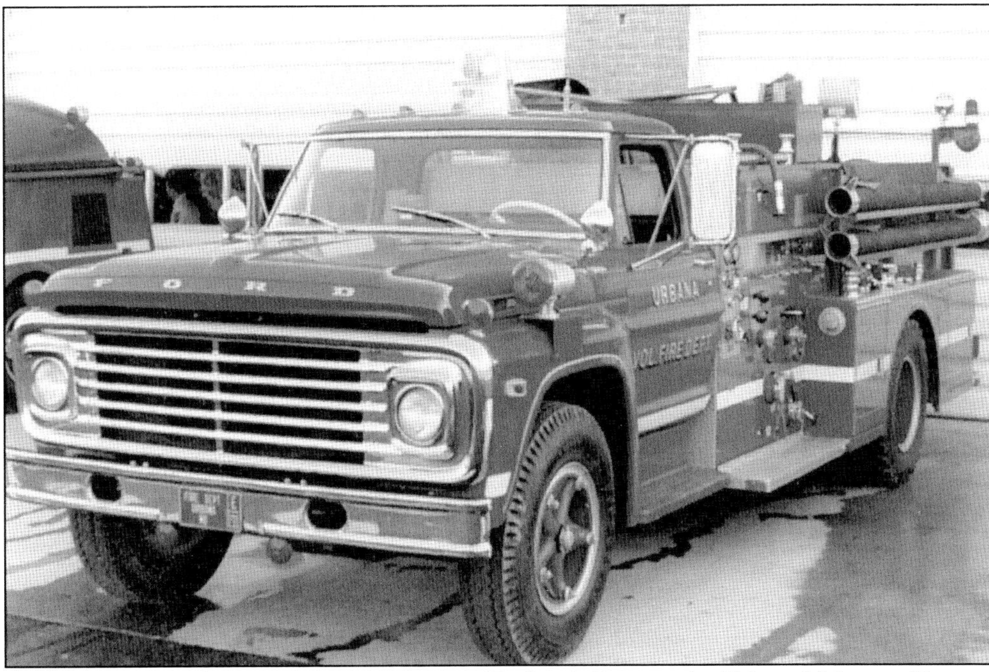

The last volunteer fire company to organize in Frederick County, the Urbana Volunteer Fire and Rescue Company actually started an ambulance service prior to the purchase of the first fire truck. This first engine was purchased from the Honeywell Corporation after limited use on a contract job site. The unit was a Ward LaFrance on a Ford chassis. (Courtesy Warren Jenkins.)

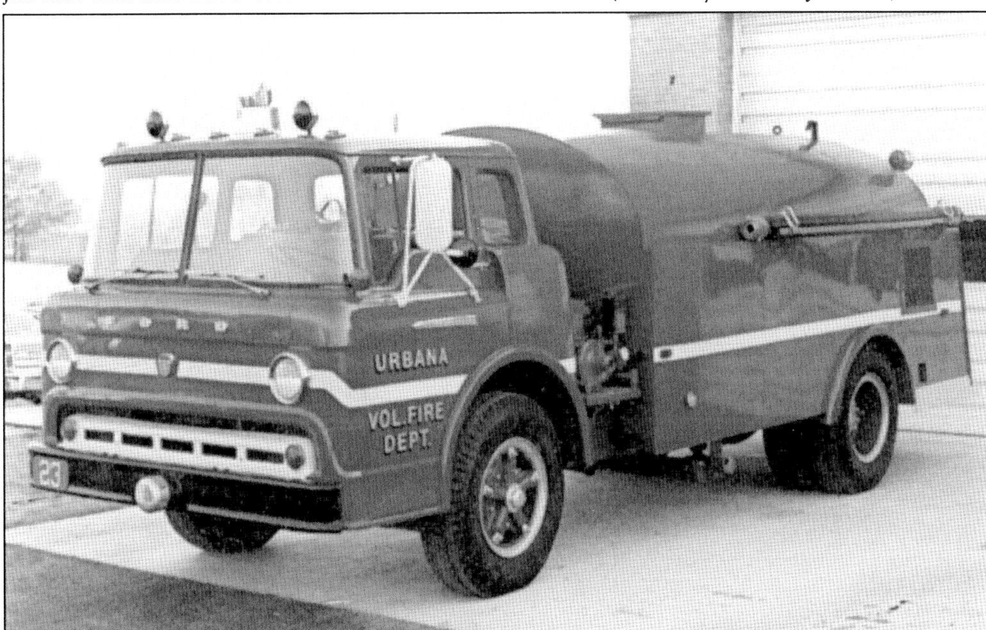

The first tanker placed in service by Urbana was originally a street sweeper purchased from the Montgomery County government. Buddy May, charter member and a chief of the department, worked for Montgomery County Vehicle Maintenance and alertly suggested the fire department purchase the vehicle for a tanker.

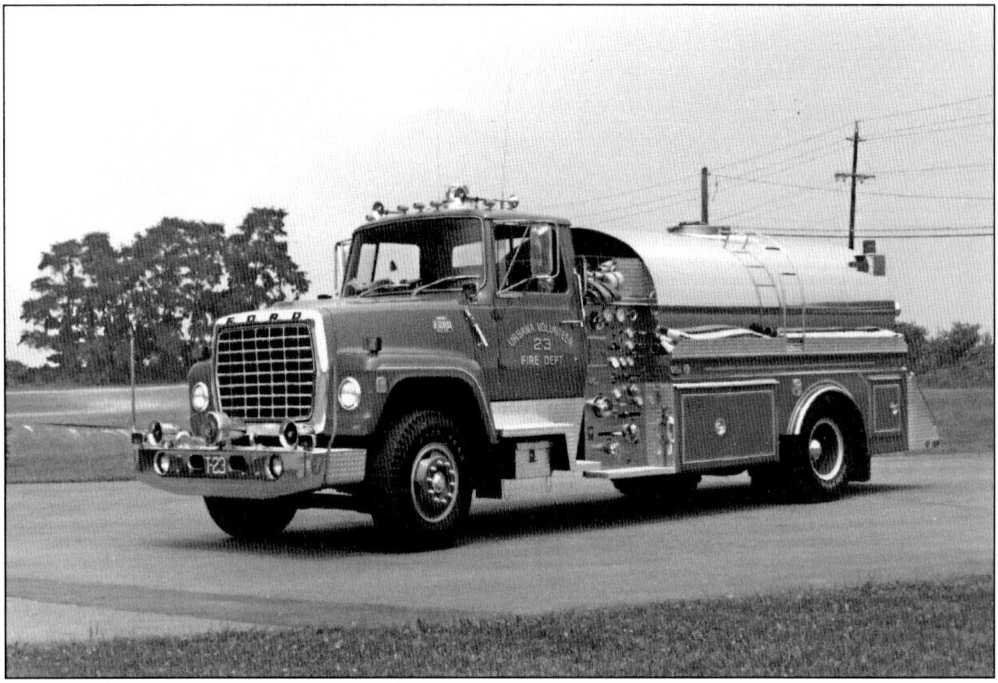

The next tanker bought was purchased new from Four Guys of Myersdale, Pennsylvania. Tanker 23 was a 1984 Ford with a 750-gallon-per-minute pump that carried 2,000 gallons of water. (Courtesy Jim May.)

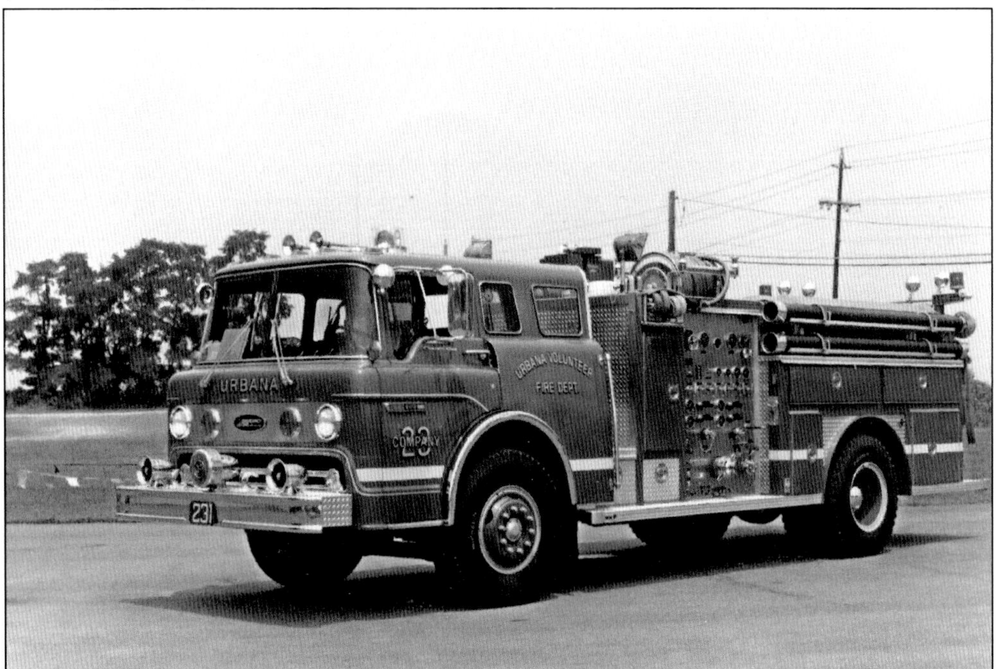

The Urbana Volunteer Fire and Rescue Company bought several E-One pumpers over the next 15 years. The first E-One was this 1980 Ford with a 1,000-gallon-per-minute pump that carried 750 gallons of water. (Courtesy Jim May.)

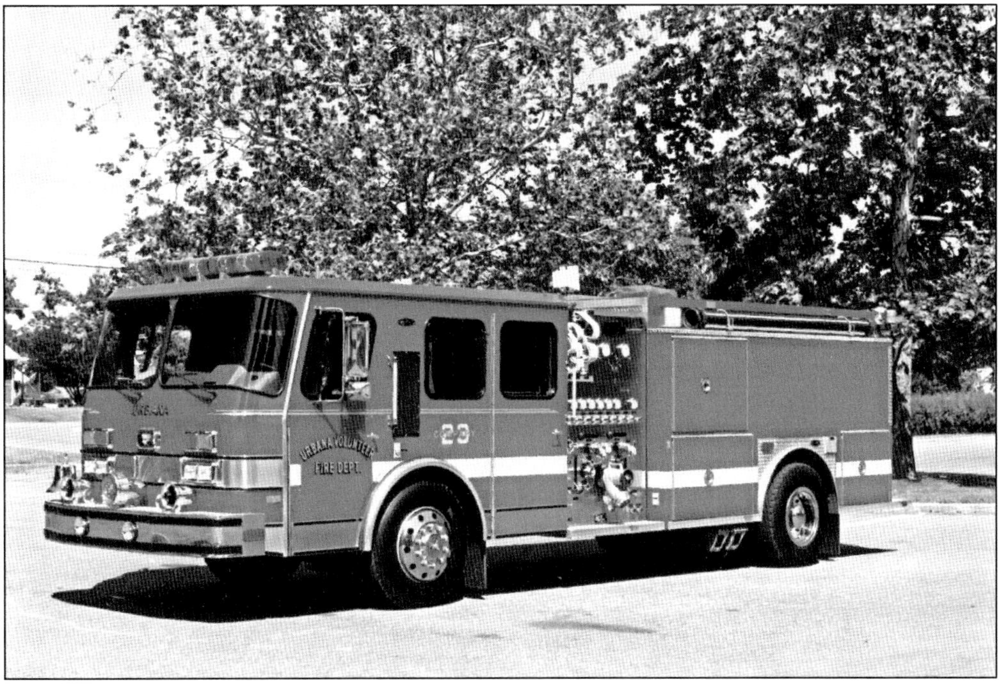

Eventually, the tankers were replaced with this 1992 E-One Cyclone engine/tanker. The engine has a 1,250-gallon-per-minute pump and carries 1,500 gallons of water. Engine Tanker 233 can be utilized as either an engine or a tanker. (Courtesy Jim May.)

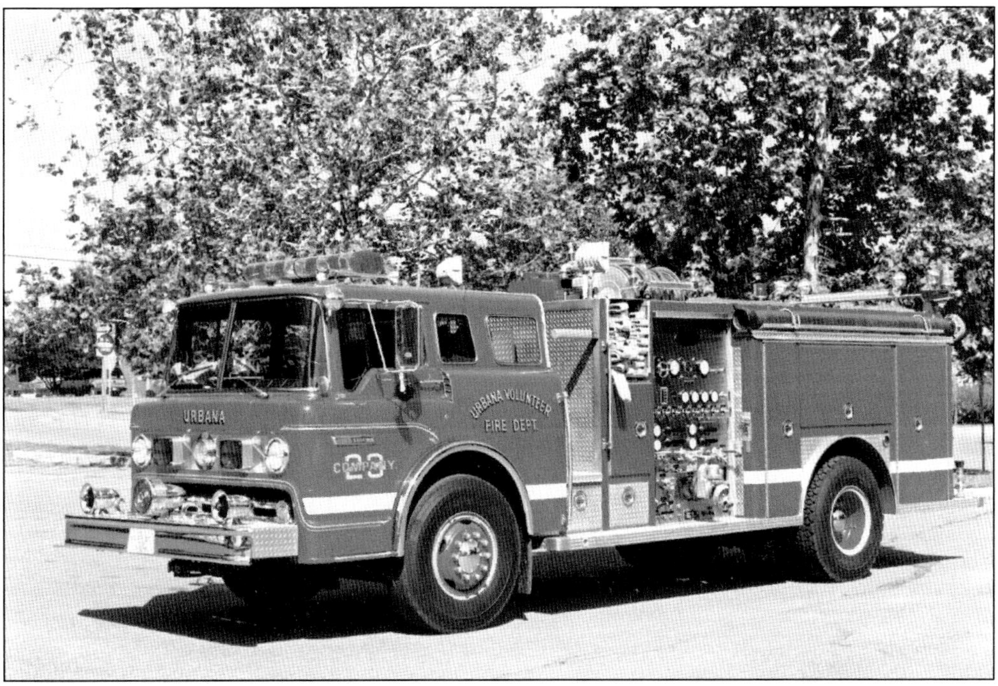

Continuing the E-One tradition, Urbana purchased this 1985 E-One pumper that included a foam proportioner as well as a 1,000-gallon-per-minute pump. The engine carried 750 gallons of water. (Courtesy Jim May.)

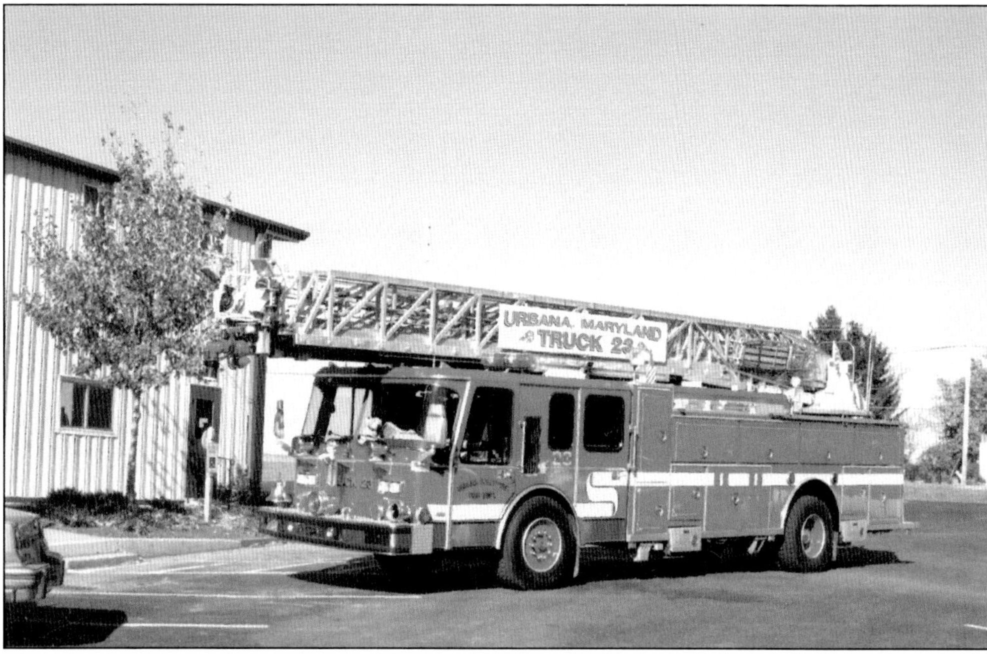

As Urbana grew, larger commercial buildings required the fire company to consider purchase of an aerial ladder truck. Truck 23 is a 1985 E-One 110-foot aerial ladder truck. The unit was previously owned by the OWL (Occoquon, Woodbridge, Lorton) Fire and Rescue in Virginia. (Courtesy Warren Jenkins.)

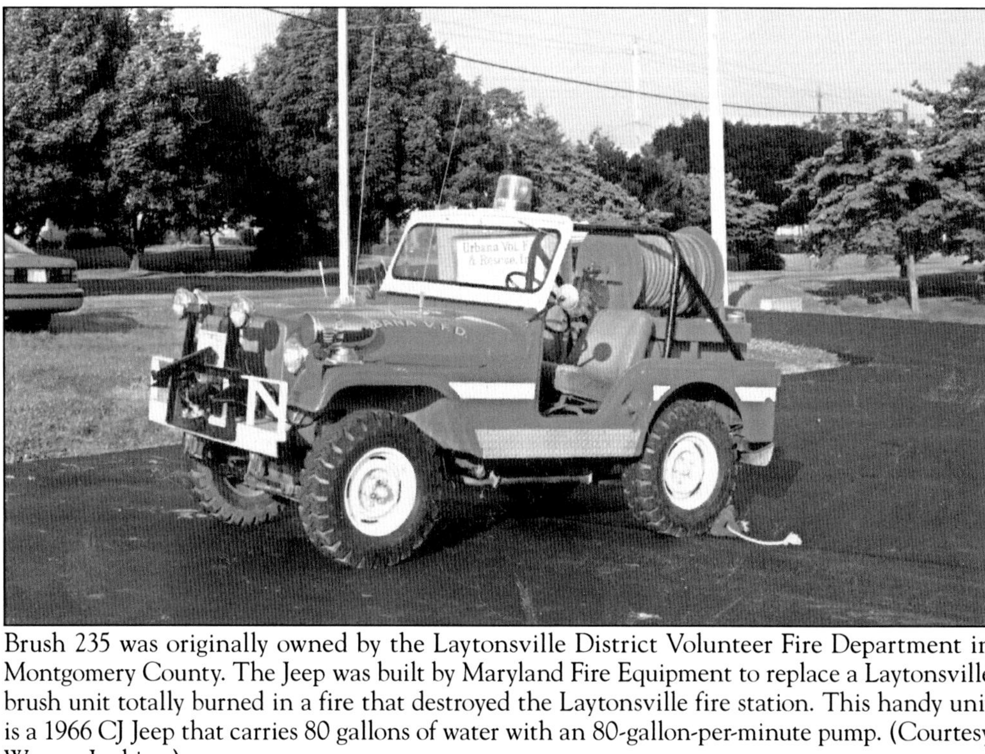

Brush 235 was originally owned by the Laytonsville District Volunteer Fire Department in Montgomery County. The Jeep was built by Maryland Fire Equipment to replace a Laytonsville brush unit totally burned in a fire that destroyed the Laytonsville fire station. This handy unit is a 1966 CJ Jeep that carries 80 gallons of water with an 80-gallon-per-minute pump. (Courtesy Warren Jenkins.)

Four

Emergency Medical Services

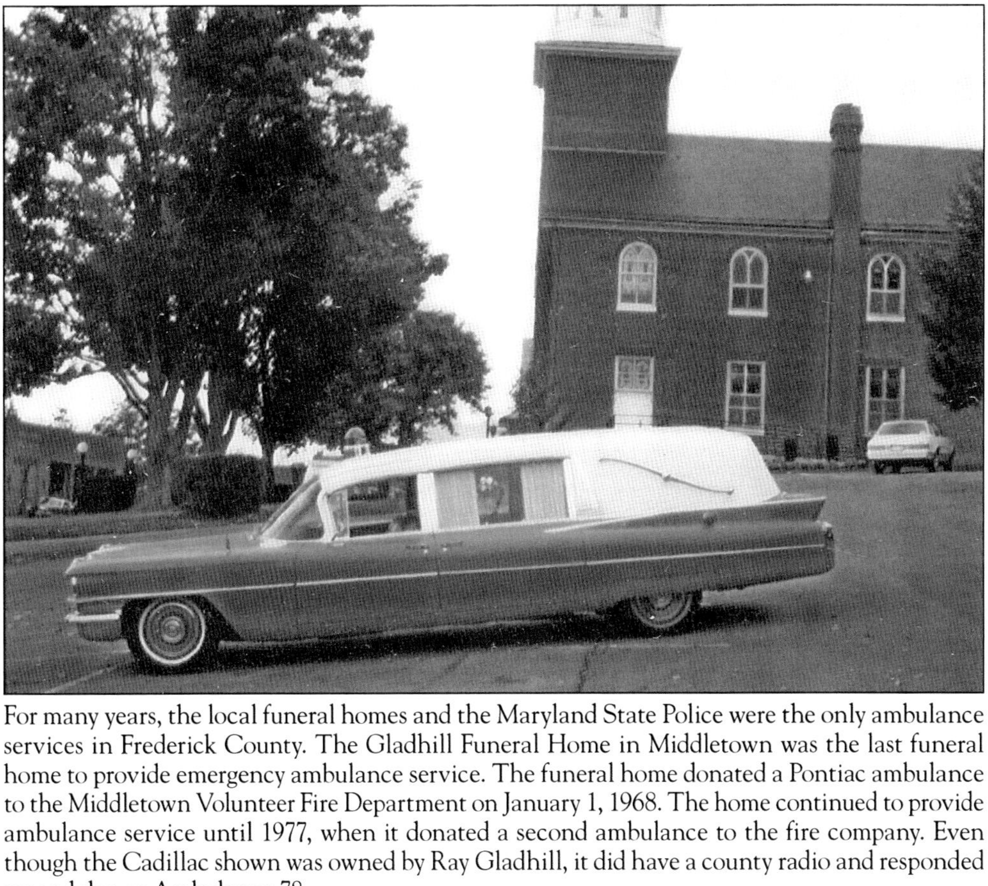

For many years, the local funeral homes and the Maryland State Police were the only ambulance services in Frederick County. The Gladhill Funeral Home in Middletown was the last funeral home to provide emergency ambulance service. The funeral home donated a Pontiac ambulance to the Middletown Volunteer Fire Department on January 1, 1968. The home continued to provide ambulance service until 1977, when it donated a second ambulance to the fire company. Even though the Cadillac shown was owned by Ray Gladhill, it did have a county radio and responded second due as Ambulance 78.

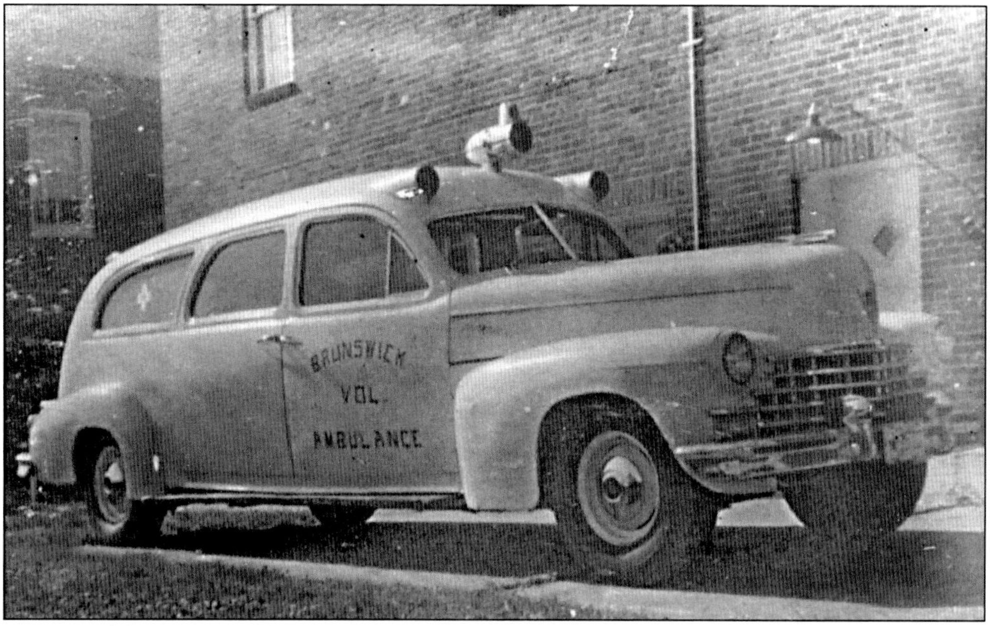

Because of the interest in forming a ground rescue unit, the Brunswick Squadron of the Civil Air Patrol was contacted to see if they were interested in purchasing this 1946 Cadillac ambulance from the Middle River Ambulance Association. Squadron commander Capt. Bernard Williams was designated to meet with Brunswick mayor James Cummings. The town voted to provide $100 towards the purchase, and on January 24, 1958, the Brunswick Volunteer Ambulance Company, Inc., was born. (Courtesy Brunswick Volunteer Ambulance Company [BVAC].)

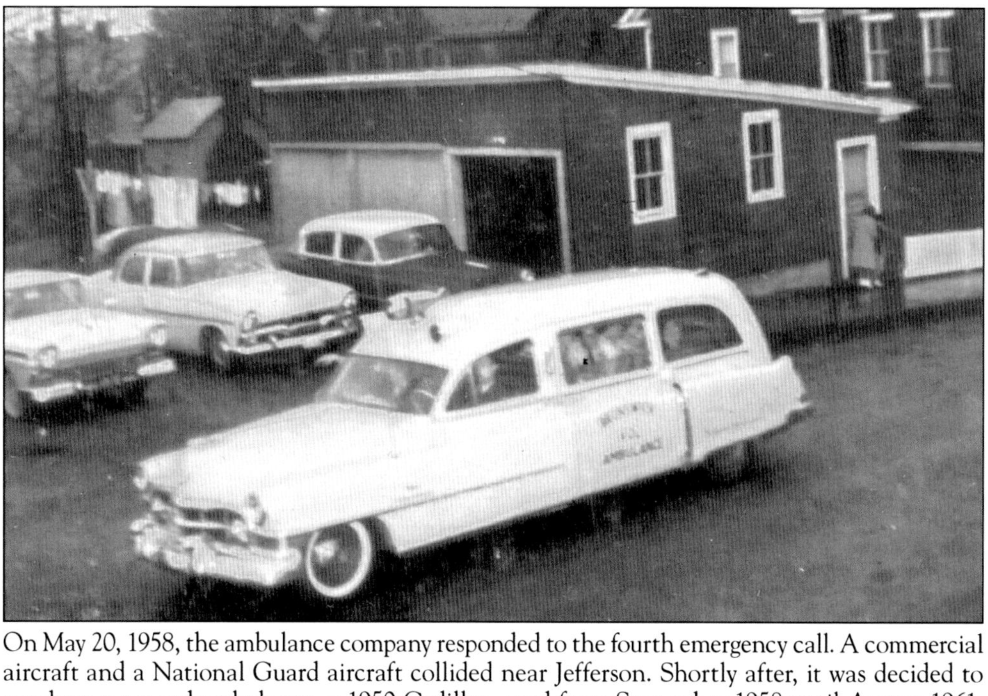

On May 20, 1958, the ambulance company responded to the fourth emergency call. A commercial aircraft and a National Guard aircraft collided near Jefferson. Shortly after, it was decided to purchase a second ambulance, a 1950 Cadillac, used from September 1958 until August 1961. The unit was designated 19A. (Courtesy BVAC.)

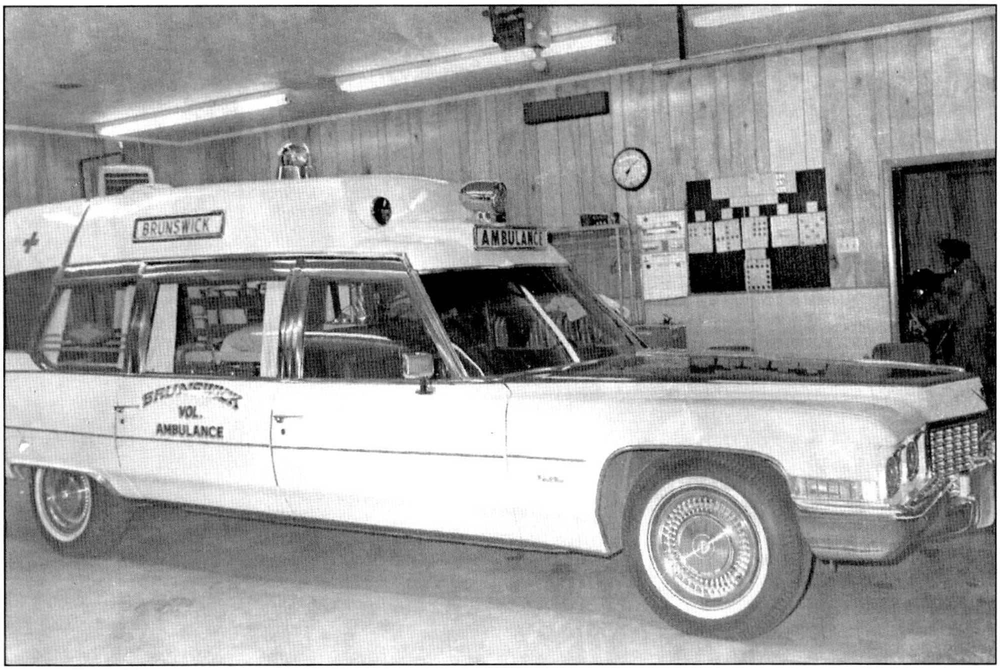

Brunswick continued to purchase Cadillac ambulances. This unit was bought in October 1969 to replace a 1961 Cadillac. (Courtesy BVAC.)

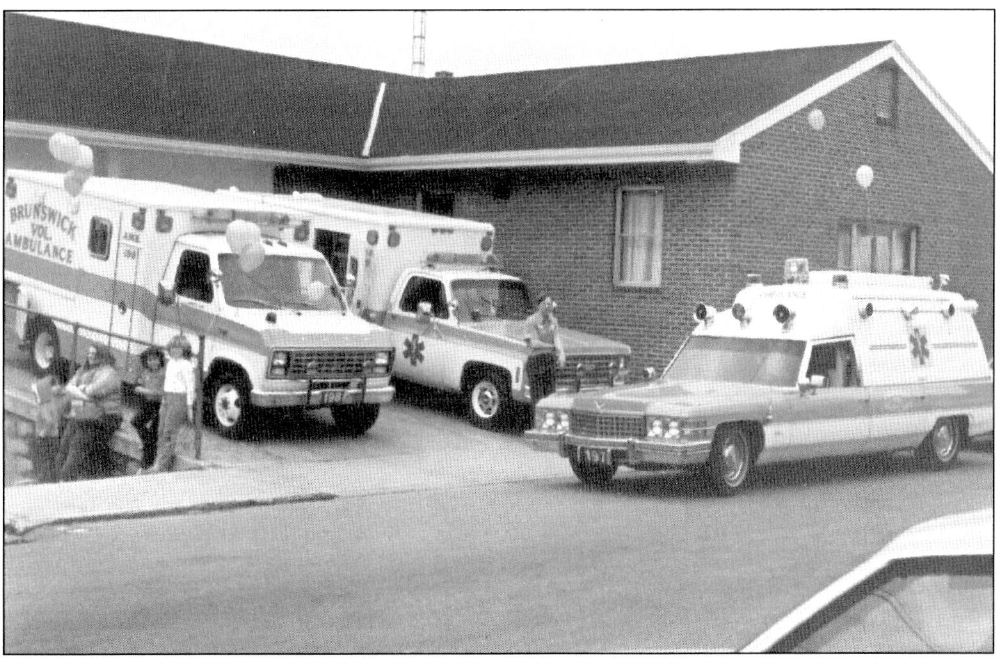

The 1974 Cadillac shown remained in service until 1986. The unit is shown in front of the ambulance building with the 1979 Ford Wheel Coach on the left and the 1977 Chevrolet Horton that replaced the 1969 Cadillac. (Courtesy BVAC.)

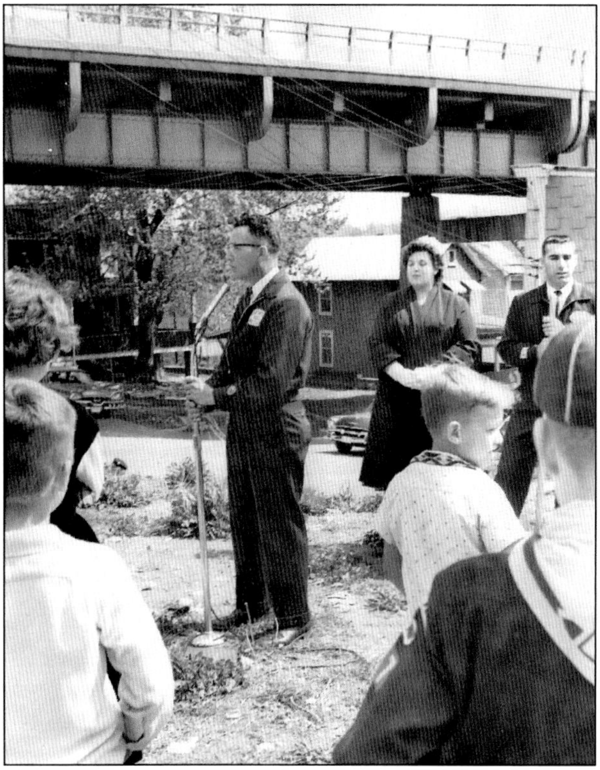

Lenwood Moss welcomes the crowd to the groundbreaking of the ambulance building at 200 West Potomac Street in May 1961. Behind him are Wanda Porter (left), Women's Auxiliary president, and Clair Ebersole (right), president of the Brunswick Volunteer Ambulance Company. (Courtesy BVAC.)

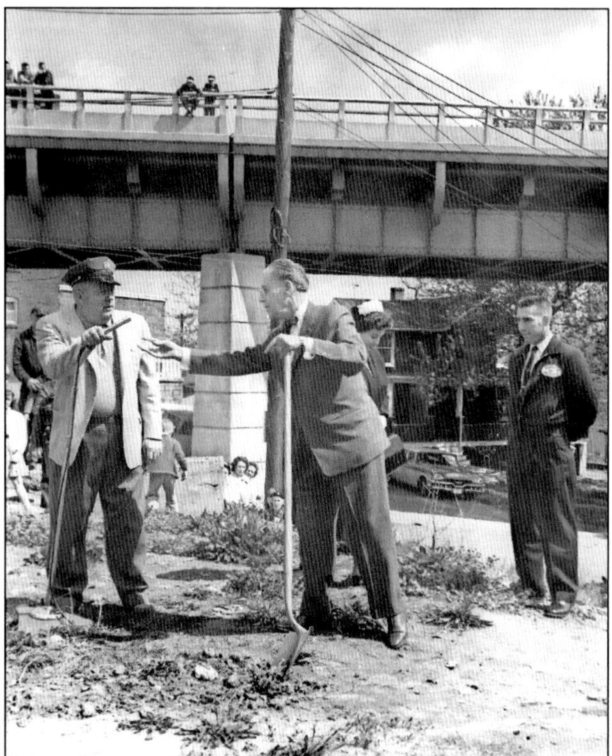

Chief H. E. "Sonny" Cannon of the Brunswick Volunteer Fire Company, serving as master of ceremonies of the groundbreaking, hands the microphone to Mayor James Cummings of Brunswick as he turns the first shovel of dirt. Clair Ebersole, president of the Brunswick Ambulance Company, looks on. (Courtesy BVAC.)

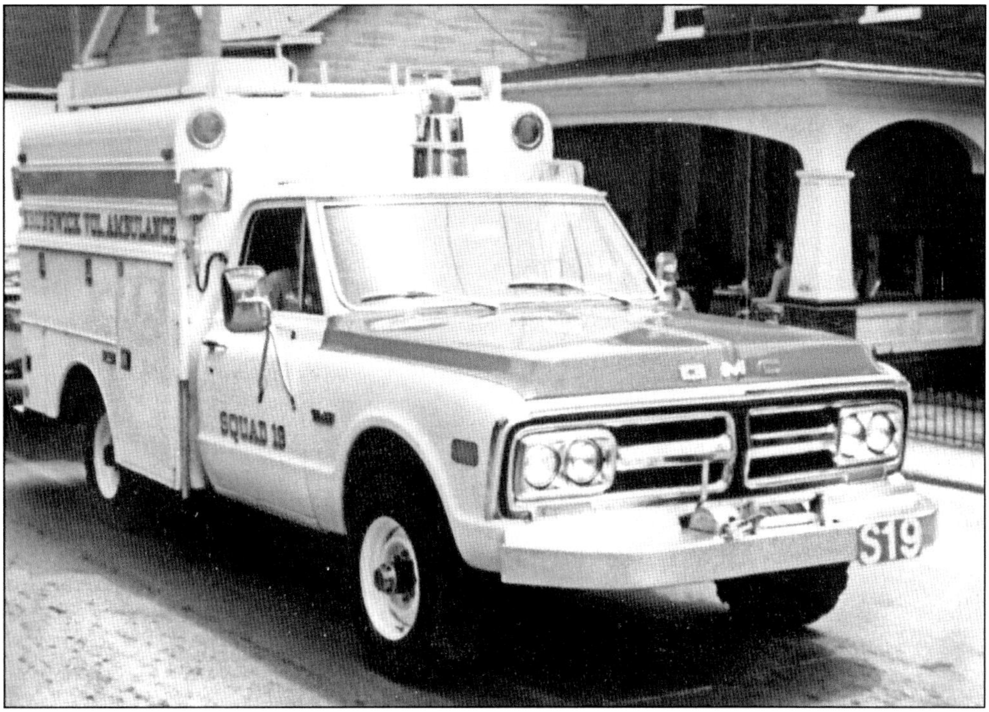

In 1971, this GMC light-duty rescue truck was placed in service to respond to auto accidents and water rescues along the river. Fifteen members completed the rescue course at the University of Maryland Fire Service Extension in October 1971. (Courtesy BVAC.)

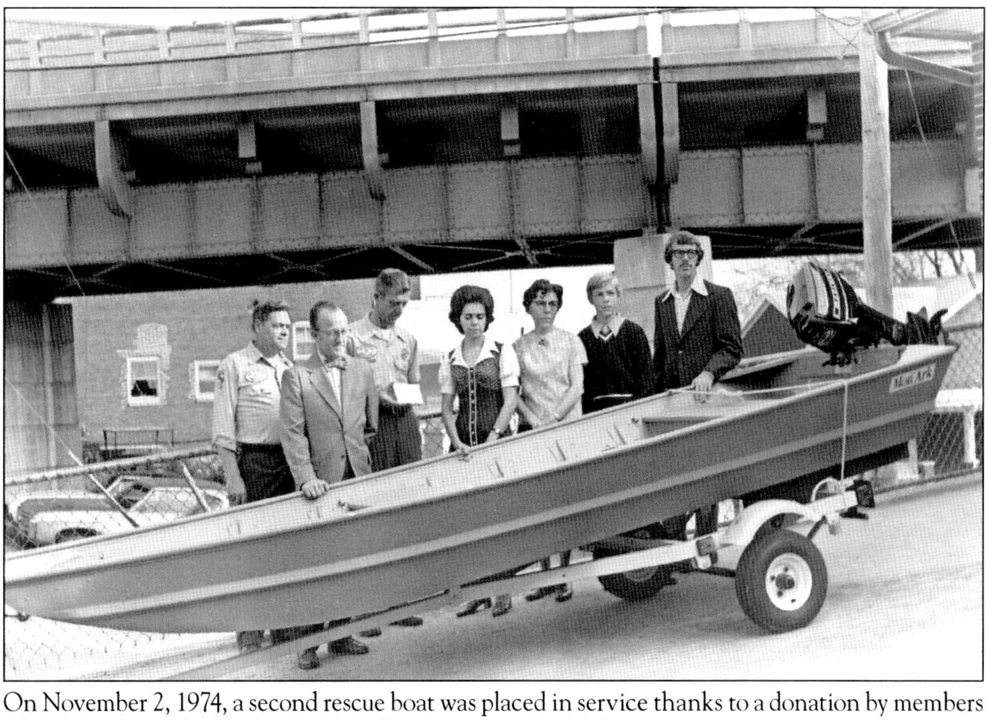

On November 2, 1974, a second rescue boat was placed in service thanks to a donation by members of the Sam Brengle family of Ijamsville. (Courtesy BVAC.)

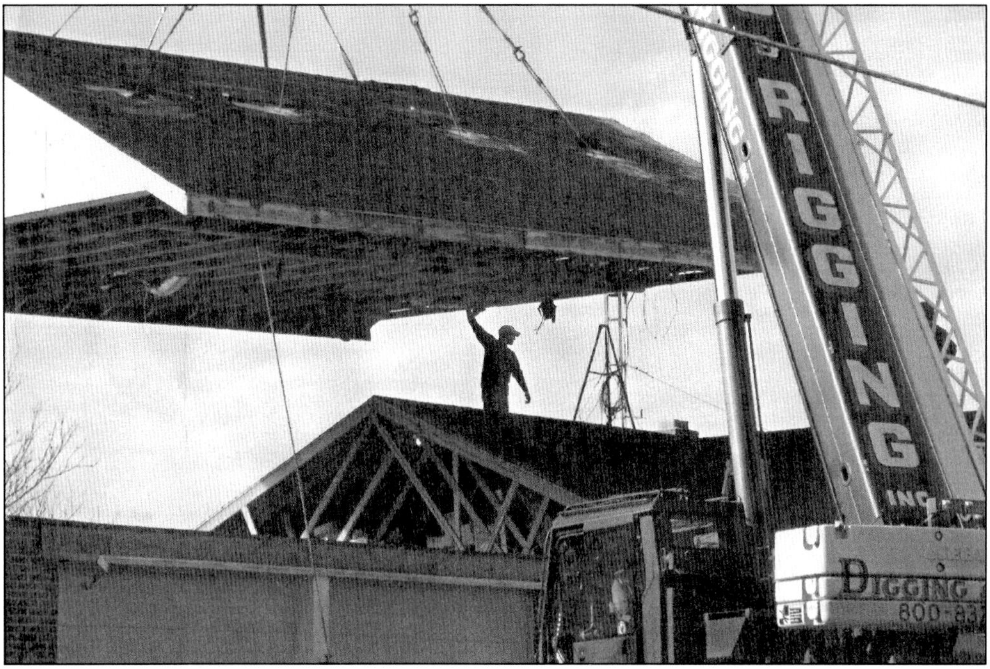

In an unusual construction project, the Brunswick Volunteer Ambulance Company hired a contractor and heavy crane from Digging and Rigging to raise the roof on the ambulance bays to accommodate today's larger apparatus. (Courtesy BVAC.)

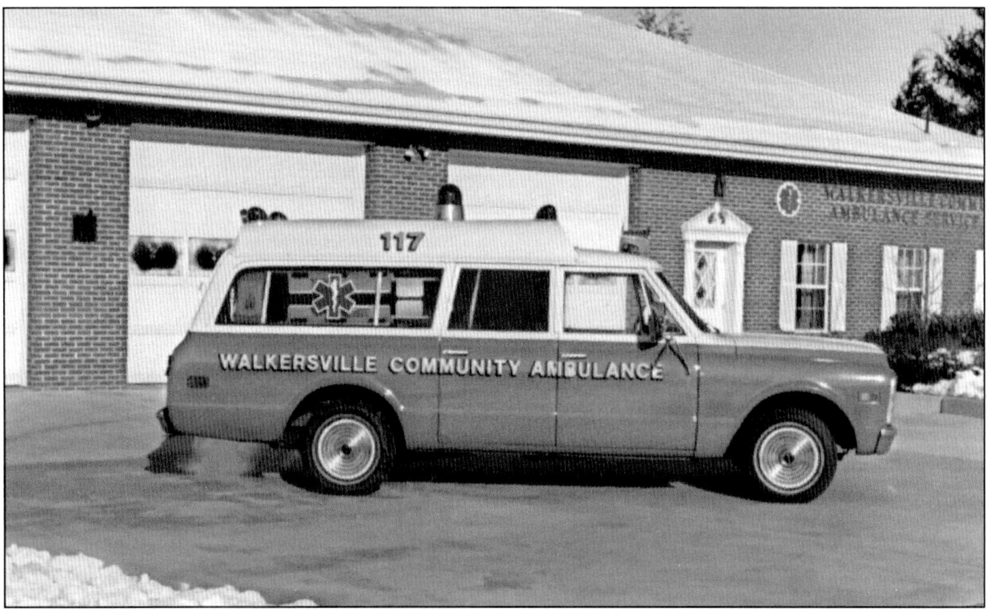

In 1977, a planning committee, in cooperation with the Walkersville Volunteer Fire Company, began to review the needs of ambulance service in the Walkersville area. The organizational meeting was held in May 1977. August 18, 1977, was the official incorporation as the Walkersville Community Ambulance Service. The first ambulance was this 1969 Chevrolet Suburban, purchased with the help of the Glade Valley Lions Club. Note the ambulance number is Ambulance 117. (Courtesy Walkersville Volunteer Rescue Company [WVRC].)

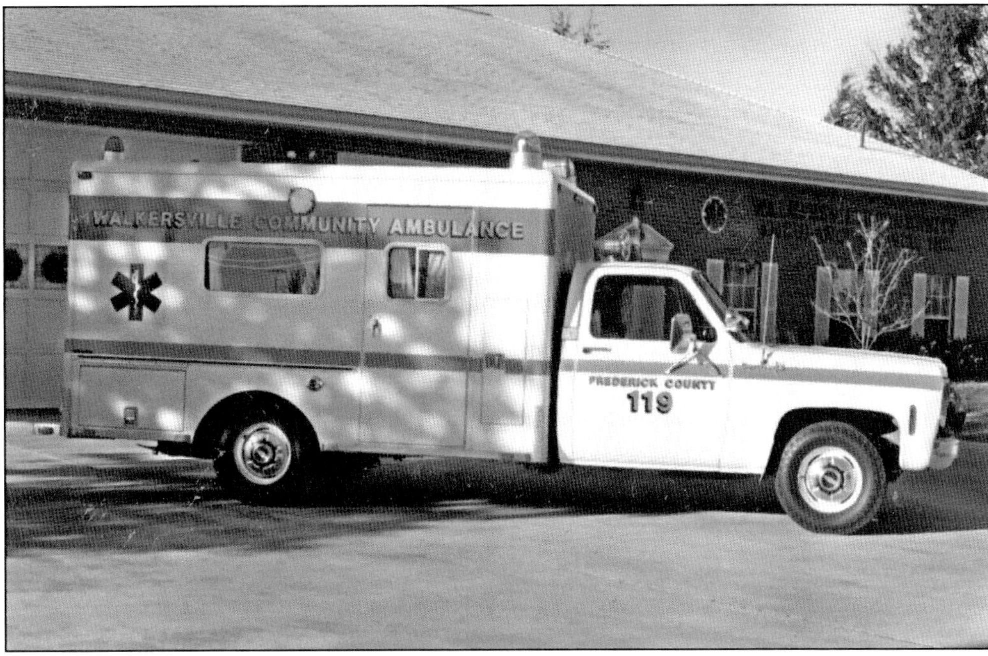

Ambulance 119 was a 1974 Ford Horton purchased from the Bethesda–Chevy Chase Rescue Squad. The gold striping was the trademark of the BCC Rescue Squad. During this time period, the ambulances responded out of the Walkersville Fire Company and used the fire company numbering series for several years. It was eventually renumbered A249. (Courtesy WVRC.)

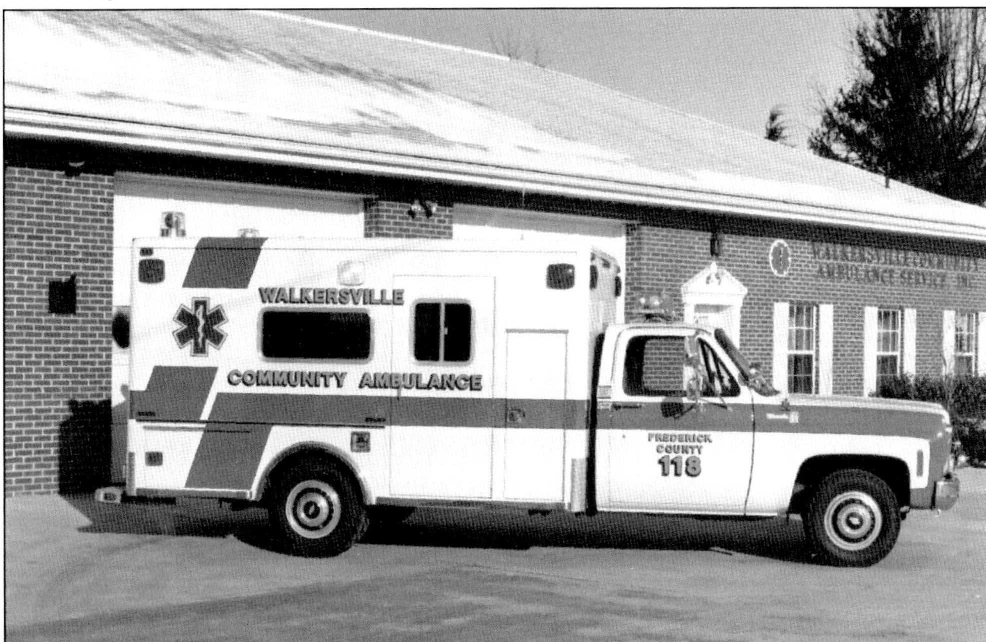

Ambulance 118, the first new ambulance purchased by the company, was a 1980 Chevrolet Horton. This unit was renumbered A248 when the company was assigned a separate number. (Courtesy WVRC.)

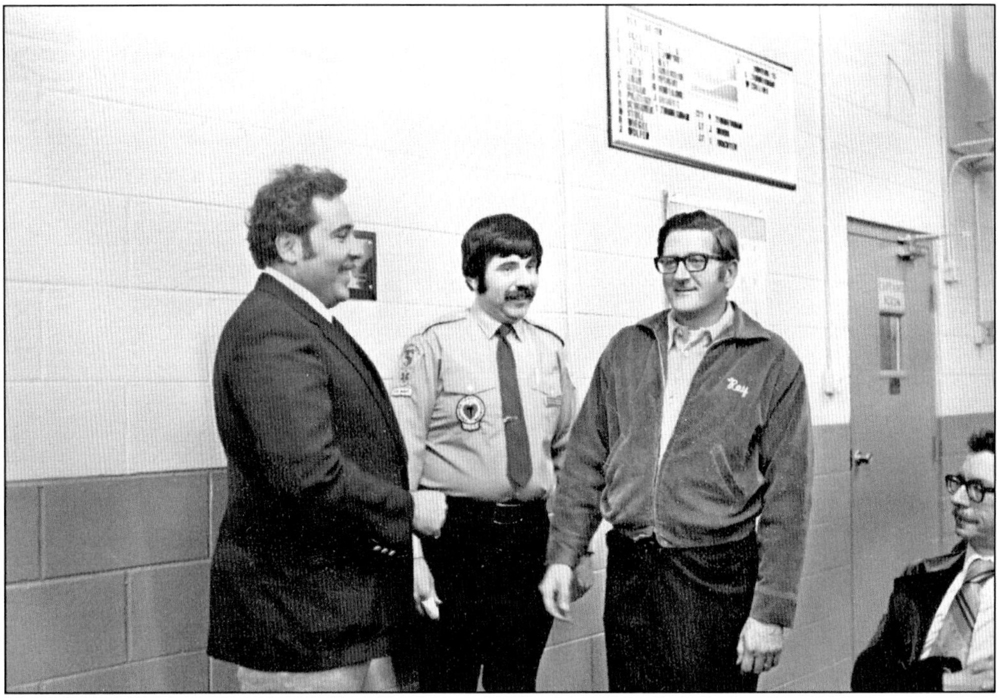

Pictured from left to right, Art Piltasky, Austin Schroyer, and Ray Dayhoff are introduced at the dedication of the ambulance building on Frederick Street in Walkersville. (Courtesy Susie Kyle.)

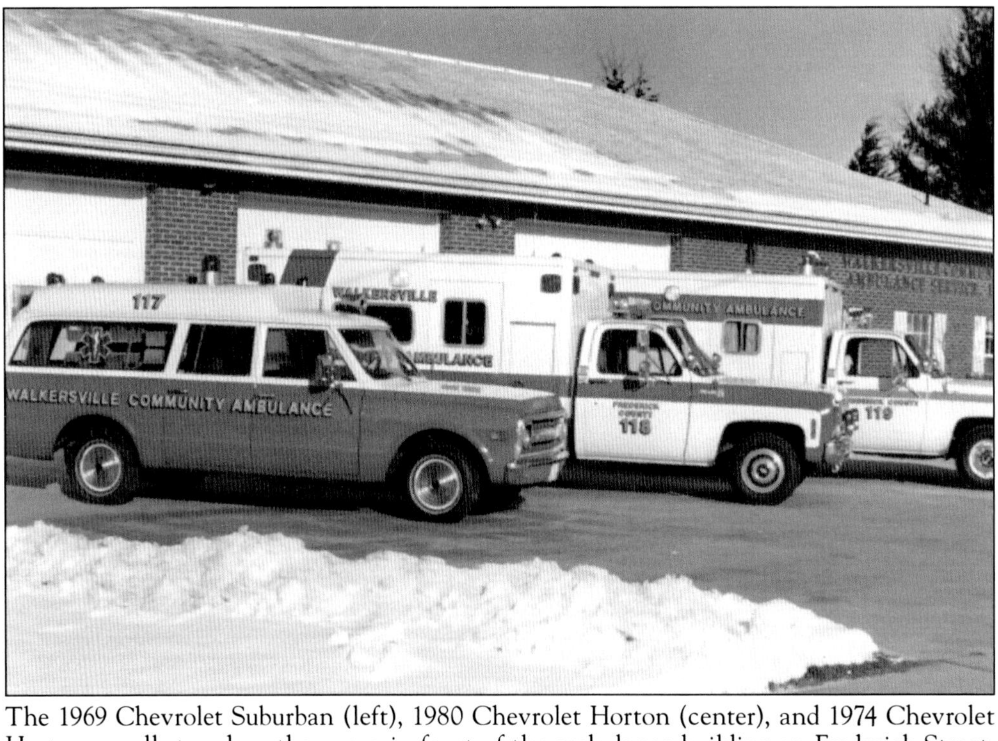

The 1969 Chevrolet Suburban (left), 1980 Chevrolet Horton (center), and 1974 Chevrolet Horton are all staged on the apron in front of the ambulance building on Frederick Street. (Courtesy WVRC.)

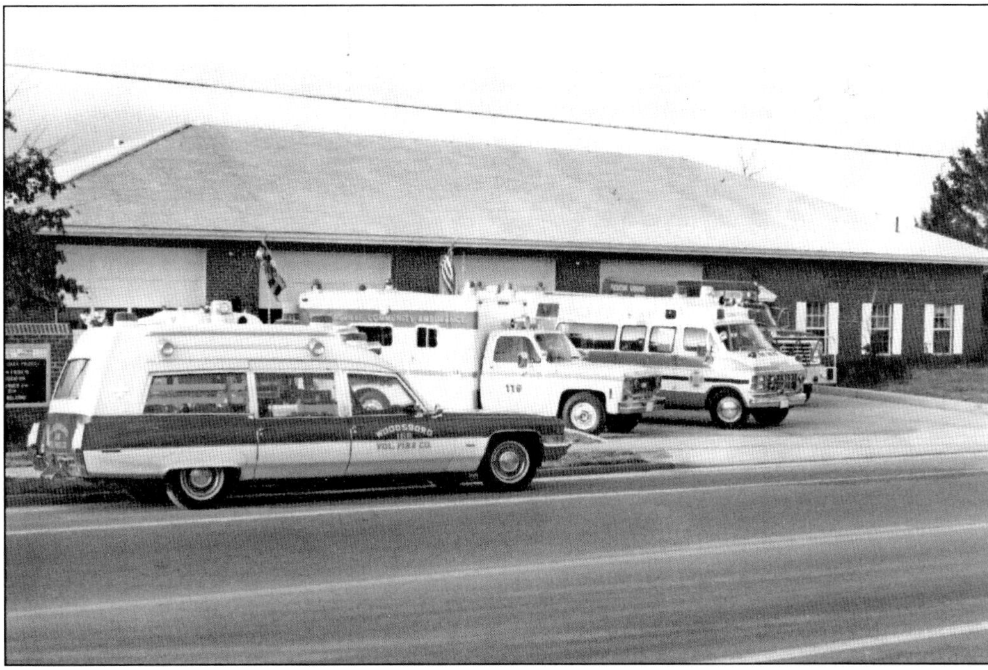

A168, the 1972 Miller/Meteor Cadillac of the Woodsboro Volunteer Fire Company, as well as A28, the Junior Fire Company's Ford Horton and Woodsboro squad truck, are posed with the A119, the 1974 Horton, at the ambulance building. (Courtesy WVRC.)

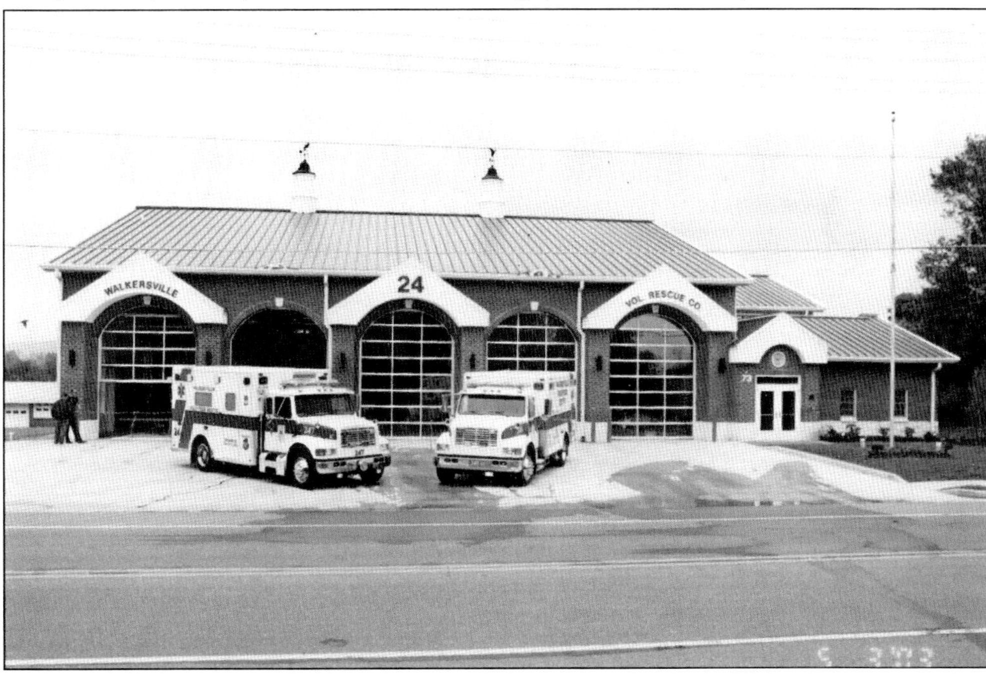

Built on the original location, the new quarters of the Walkersville Volunteer Rescue Company provide a much larger facility and room for future expansion. The administration area is two stories, while the apparatus bays offer much higher ceilings than the former quarters. (Courtesy WVRC.)

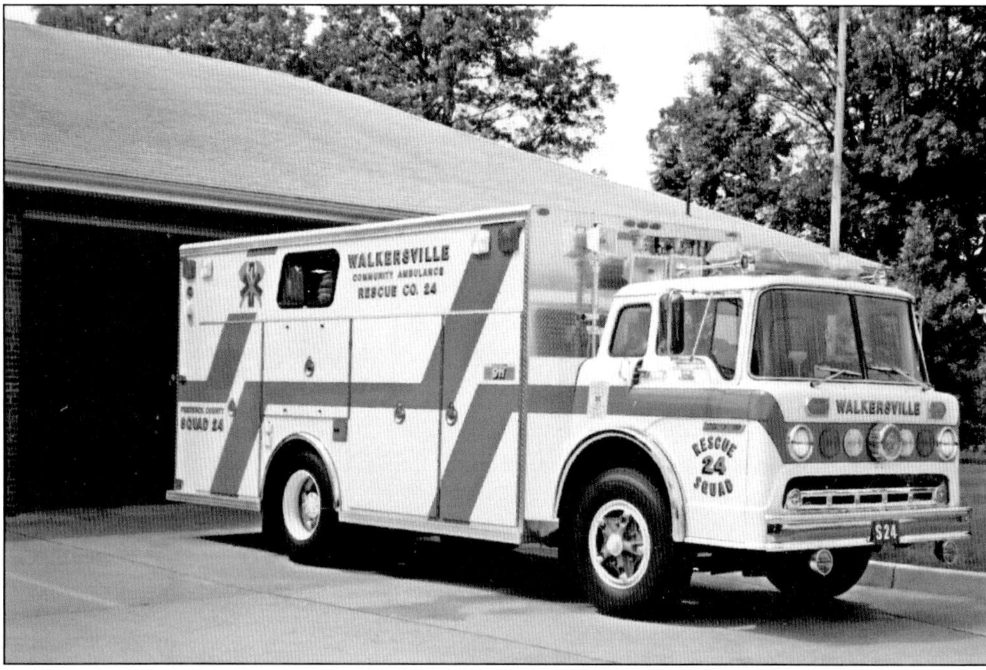

Recognizing a need for a rescue truck, the Walkersville Community Ambulance Service received a grant from the Emergency Assistance Trust Fund to purchase the 1983 Ford/EVF heavy rescue truck. This unit responded as Squad 24. (Courtesy Warren Jenkins.)

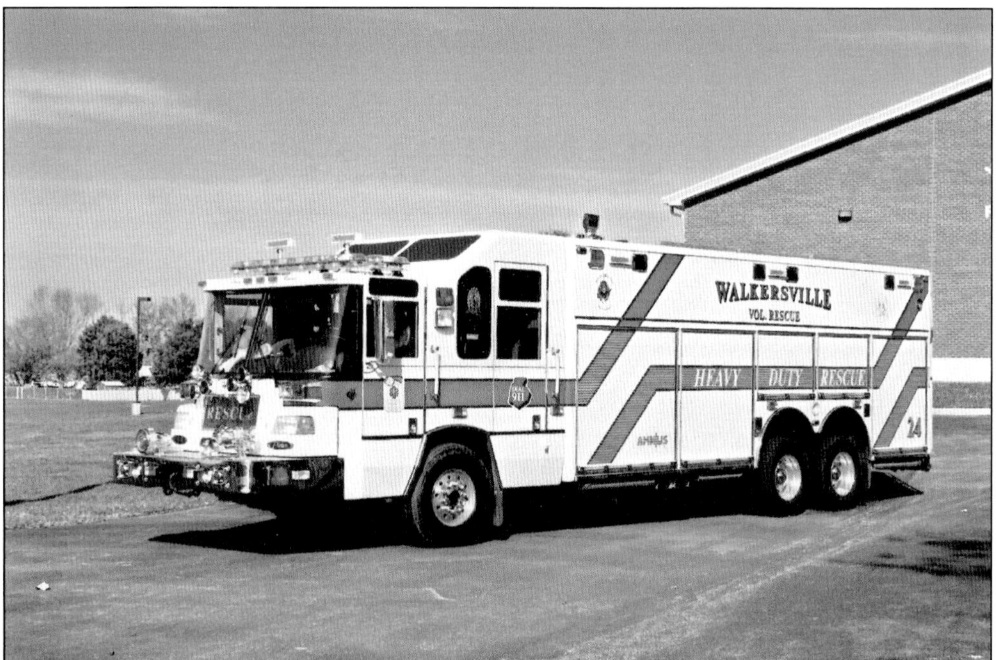

The rescue company purchased this 2001 Pierce Quantum Heavy Duty Rescue Squad to replace the Ford. This unit meets all heavy squad standards and can run multiple rescue tools at one time. (Courtesy Warren Jenkins.)

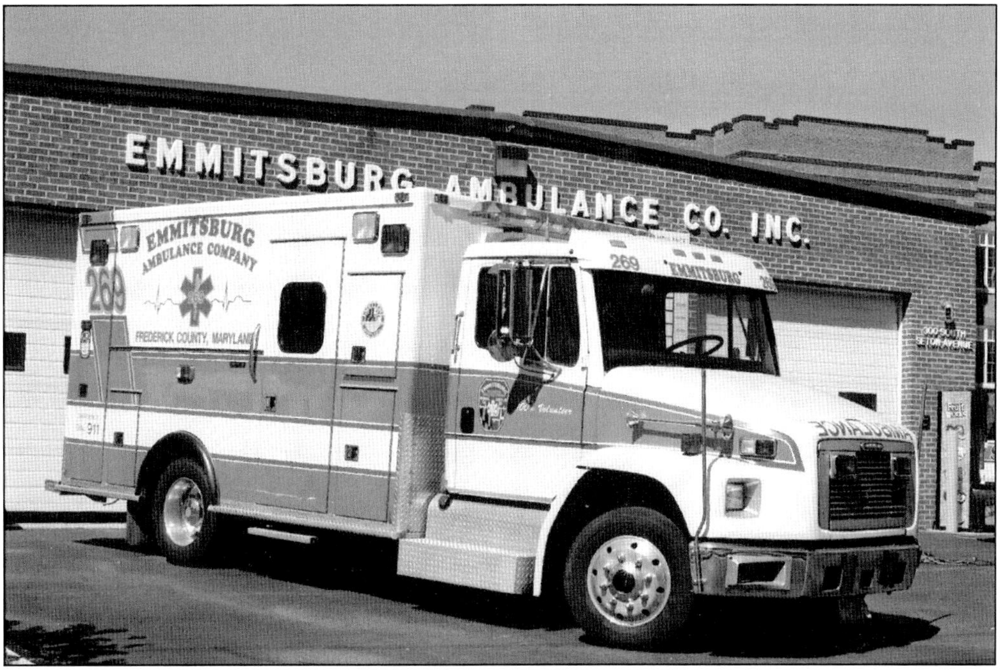

Ambulance service has been provided to the Emmitsburg area for over 50 years. For many years, the local VFW Post provided the ambulance service. Later, the American Legion also purchased an ambulance. The Emmitsburg Ambulance Company operates two ambulances and a special unit from the quarters on South Seaton Avenue. Ambulance 269 is a Horton ambulance on a Freightliner chassis. (Courtesy Warren Jenkins.)

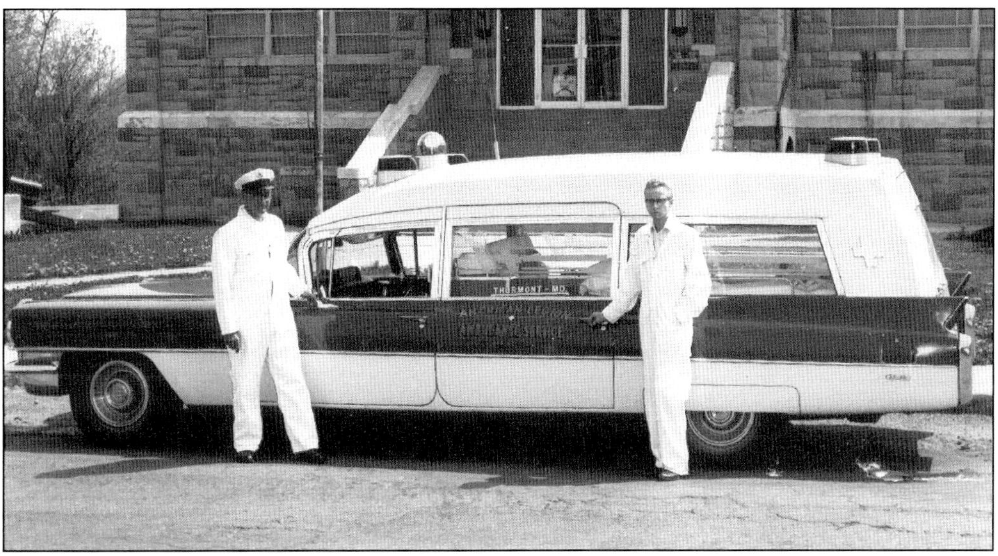

Herman Shook (left) and Terry Shook in their ambulance attendants' white jumpsuits are beside a Thurmont American Legion Cadillac ambulance. The members of the Edwin C. Creeger Post #168 of the American Legion met in July 1953 to begin discussion of an ambulance service. Prior to the Legion, ambulance service was provided by the Creager Funeral Home with a hearse that could double as an ambulance when the red light was placed on top. (Courtesy Thurmont Community Ambulance Service [TCAS].)

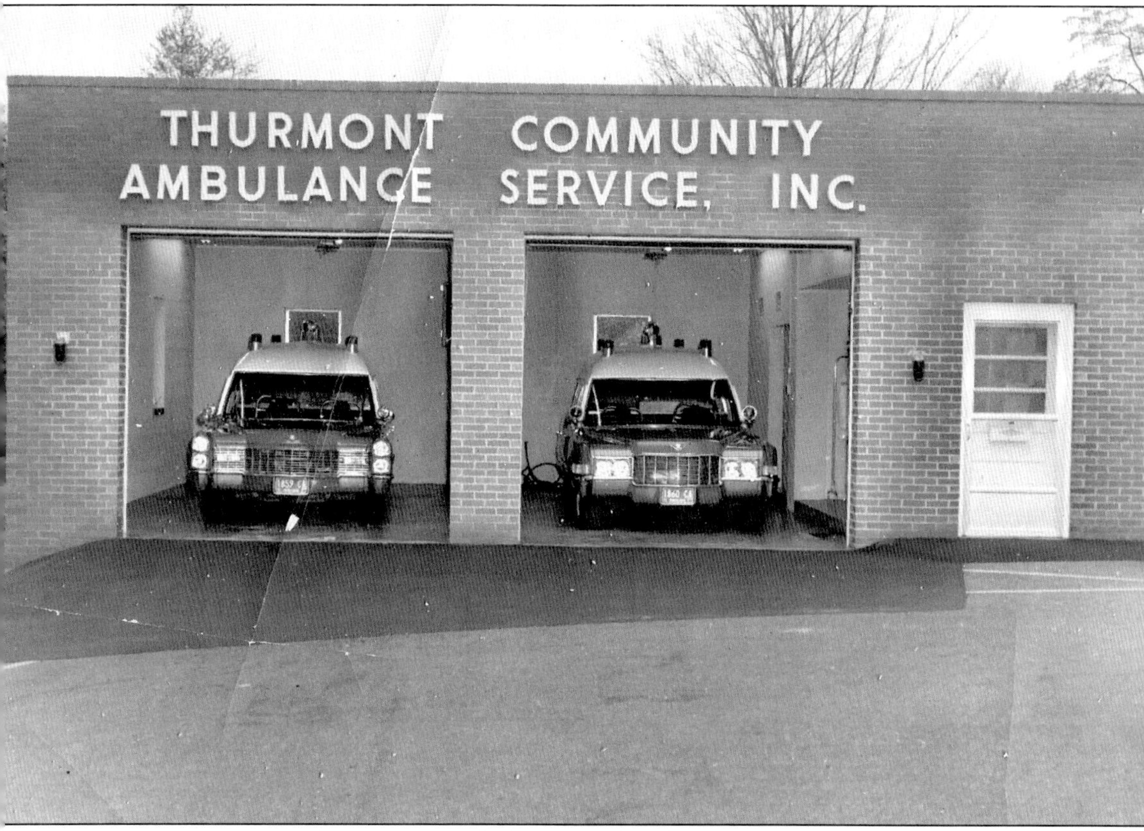

For years, the American Legion housed the ambulance in a garage below the legion hall. The Thurmont Community Ambulance Service was also housed in a building owned by the Guardian Hose Company on the rear of the parking lot. (Courtesy TCAS.)

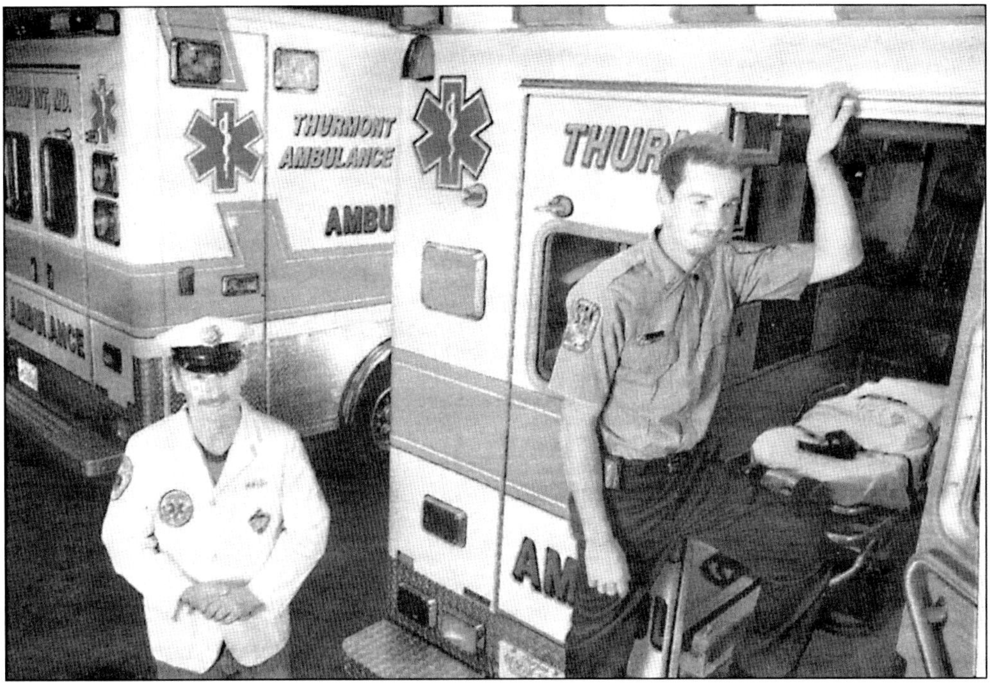

Wayne Stackhouse, left, wears the original uniform and hat worn by the personnel of the Thurmont Community Ambulance Service. D. J. Ott, standing in the ambulance, wears the uniform used today. (Courtesy TCAS.)

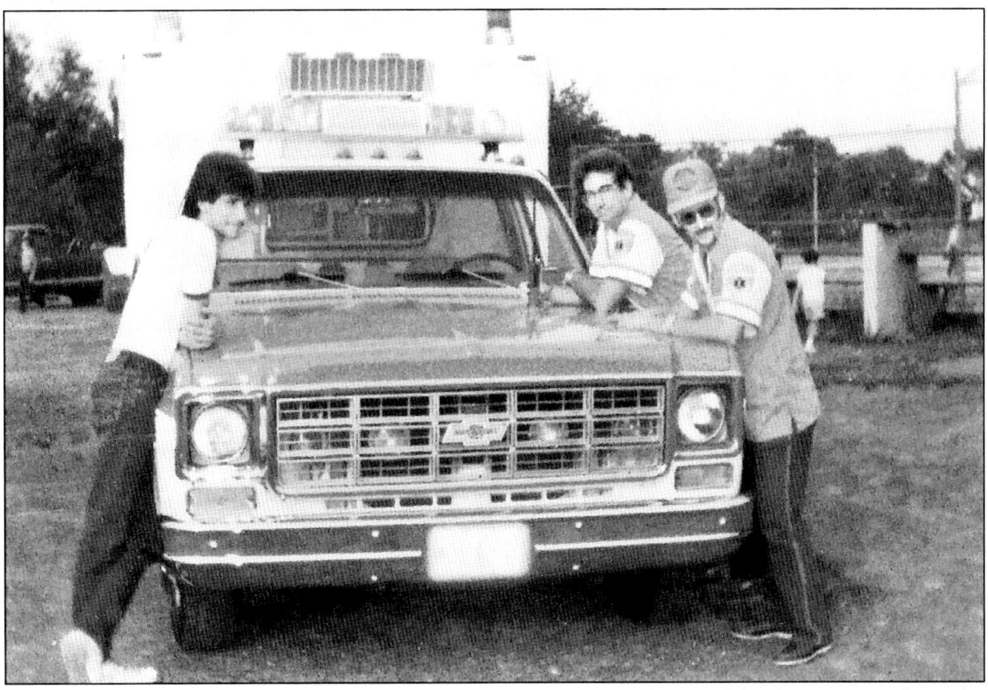

Members of the Thurmont Community Ambulance Service stand around the first Type 1 Chevrolet ambulance purchased to replace a Cadillac. Shown at the 1975 Chevrolet Swab are, from left to right, unidentified, Bob Oxendale, and Wayne Stackhouse. (Courtesy TCAS.)

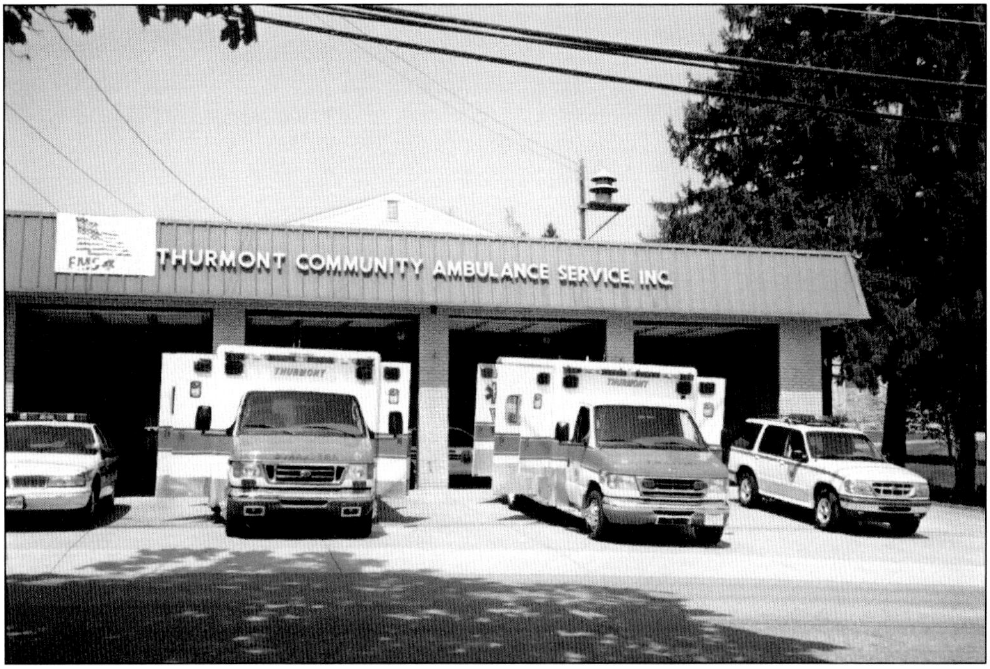

The ambulance company purchased the church next to the Guardian Hose Company and built an apparatus bay on the front of the structure. On the left, Medic 30, a county Advanced Life Support Unit, is now based in the station. In addition to the ambulances, the duty vehicle is shown on the right. (Courtesy Warren Jenkins.)

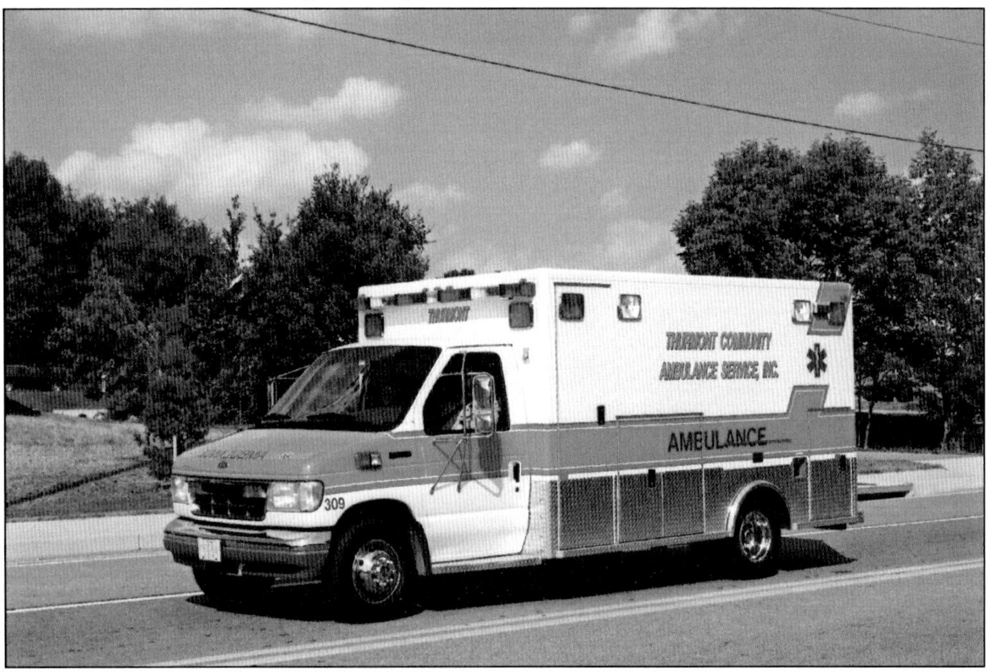

Ambulance 309 is a 2001 Ford Horton and part of the modern fleet of ambulances maintained today. The company now maintains three ambulances in service as well as a duty vehicle. (Courtesy Warren Jenkins.)

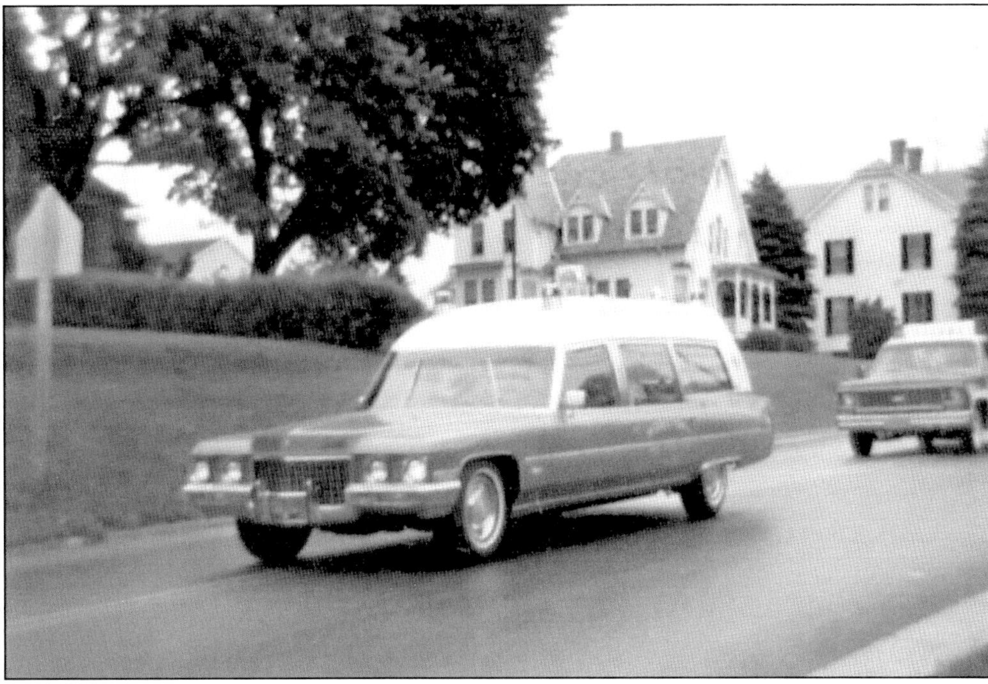

The first ambulance purchased by the Middletown Volunteer Fire Company was this 1971 S&S Cadillac ambulance. The first ambulances were previously owned by the Gladhill Funeral Home in Middletown. This ambulance was purchased for $14,991.75. (Courtesy MVFC.)

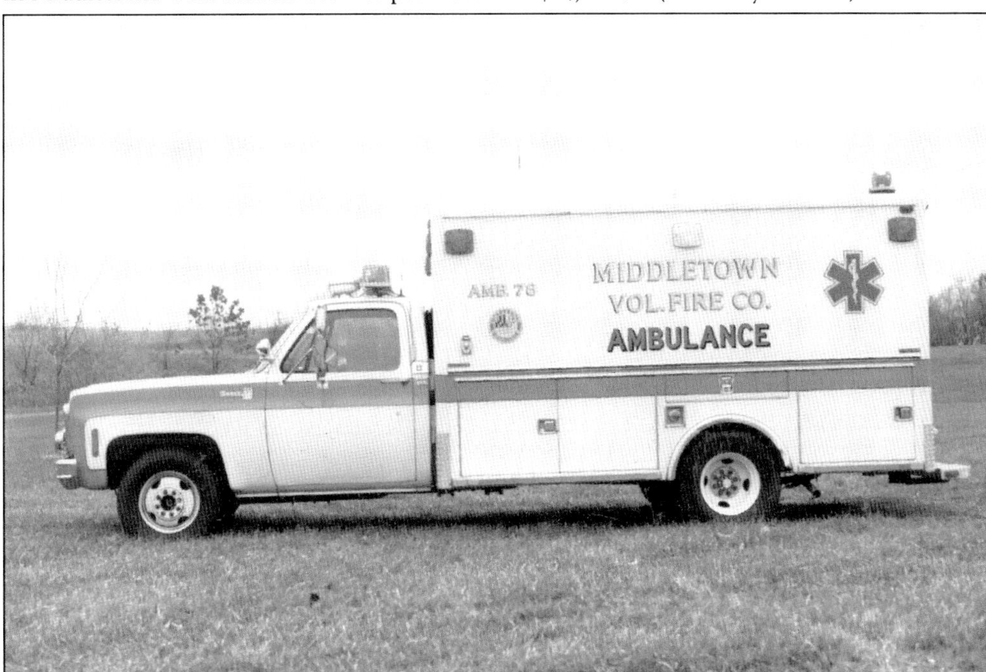

In 1977, the Middletown Volunteer Fire Company continued to improve service to the community with the purchase of this 1978 Horton ambulance on a Chevrolet chassis. This unit responded as Ambulance 78. (Courtesy MVFC.)

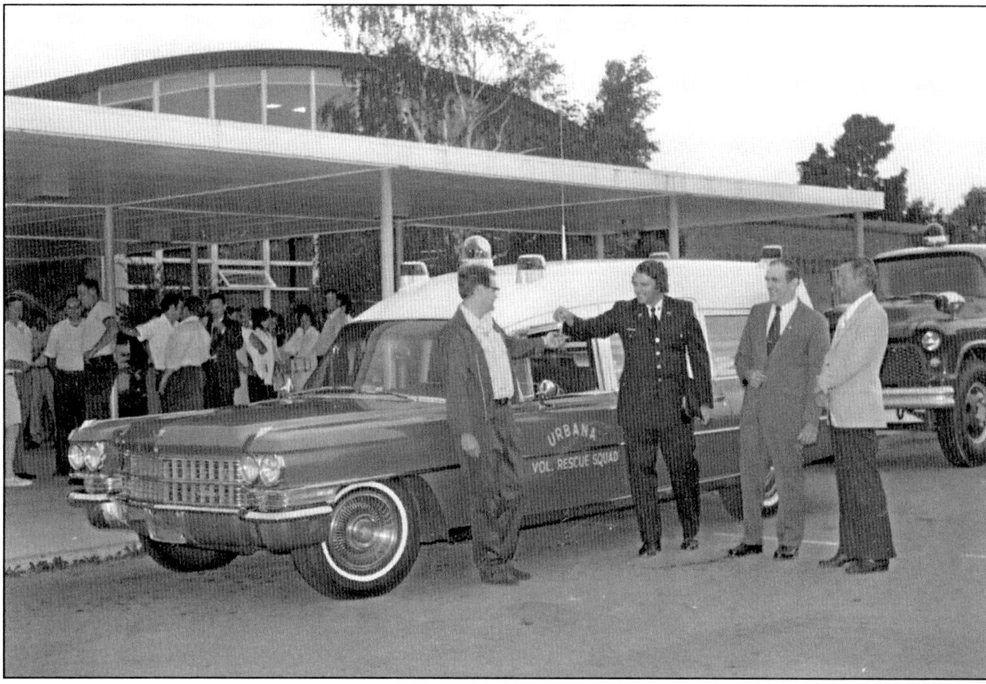

When the Urbana Volunteer Fire and Rescue Company first organized, only ambulance service was provided. The first unit was this 1963 Cadillac received from Fifth District Volunteer Fire Company, Clarksville, in Howard County. (Courtesy Susie Kyle.)

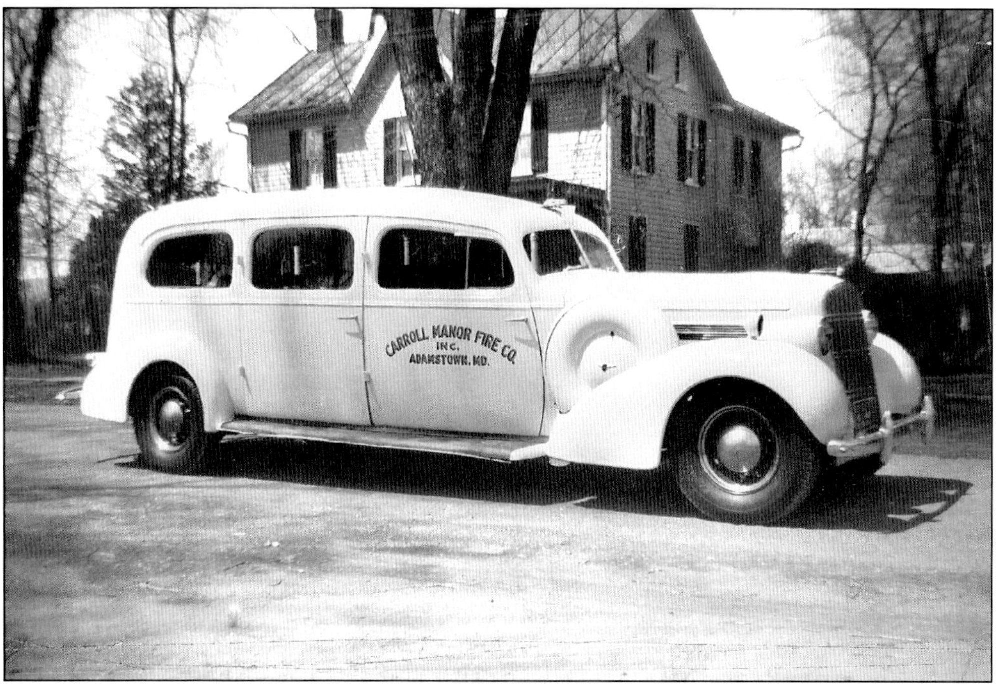

The first ambulance owned by the Carroll Manor Fire Company was this 1938 Oldsmobile Hennessey. It is believed this ambulance was previously owned by a local funeral home before being acquired by the fire company. (Courtesy CMFC.)

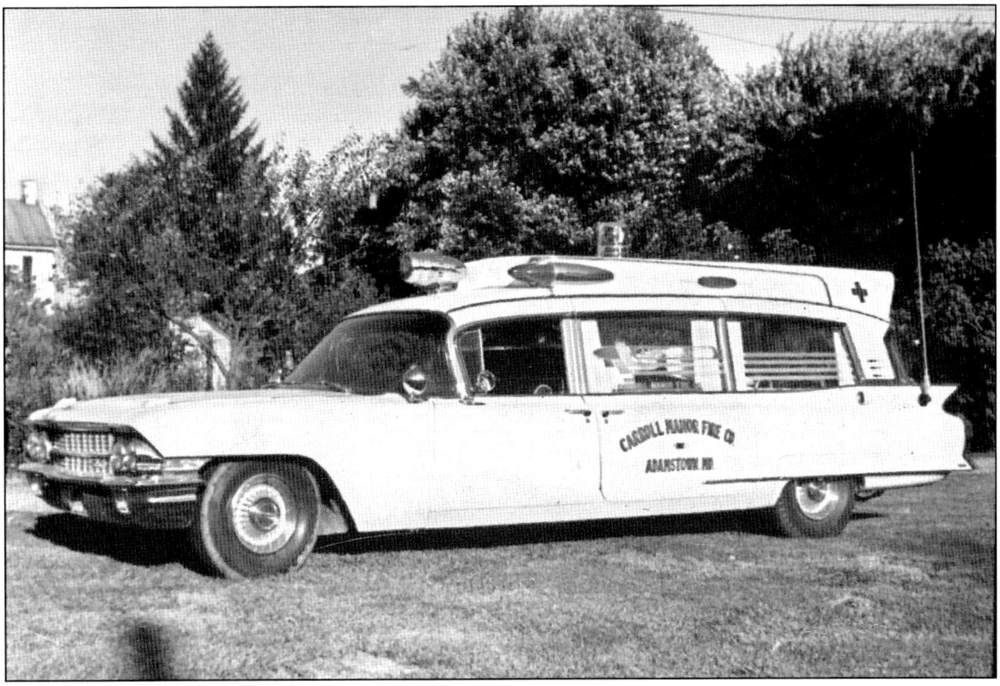

This 1958 Cadillac was purchased new by the fire company and responded to hundreds of calls during its years of service. The unique bullet-style lights were known as tunnel lights and were common on many Cadillac ambulance bodies. (Courtesy CMFC.)

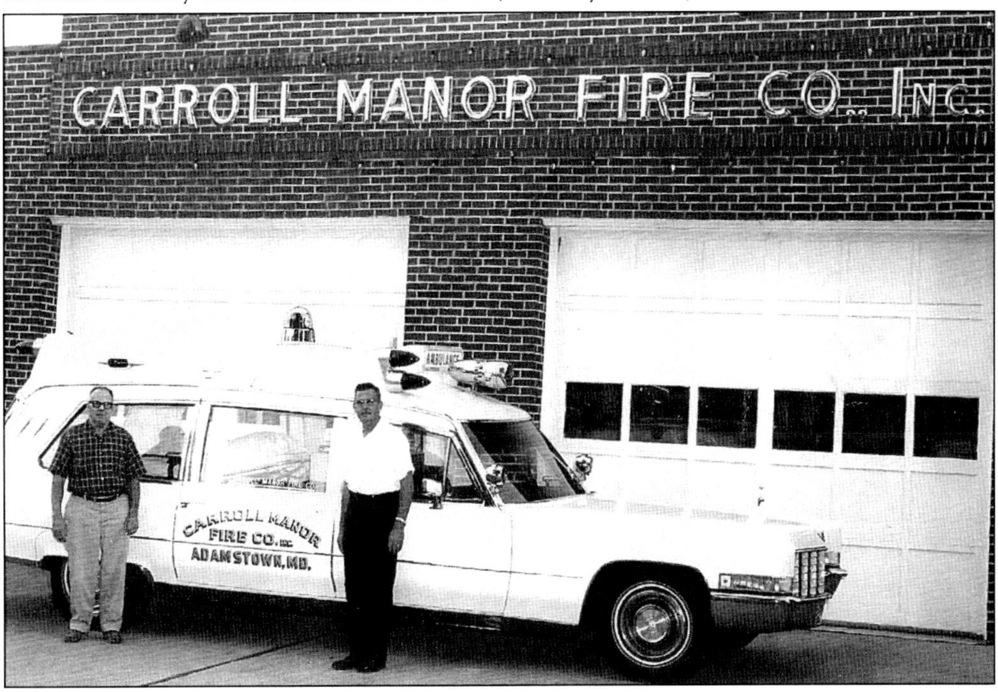

Carroll Manor continued to purchase the smooth-riding Cadillac ambulances and placed in service this 1967 ambulance. The company had two Cadillac ambulances in service well into the 1970s. (Courtesy CMFC.)

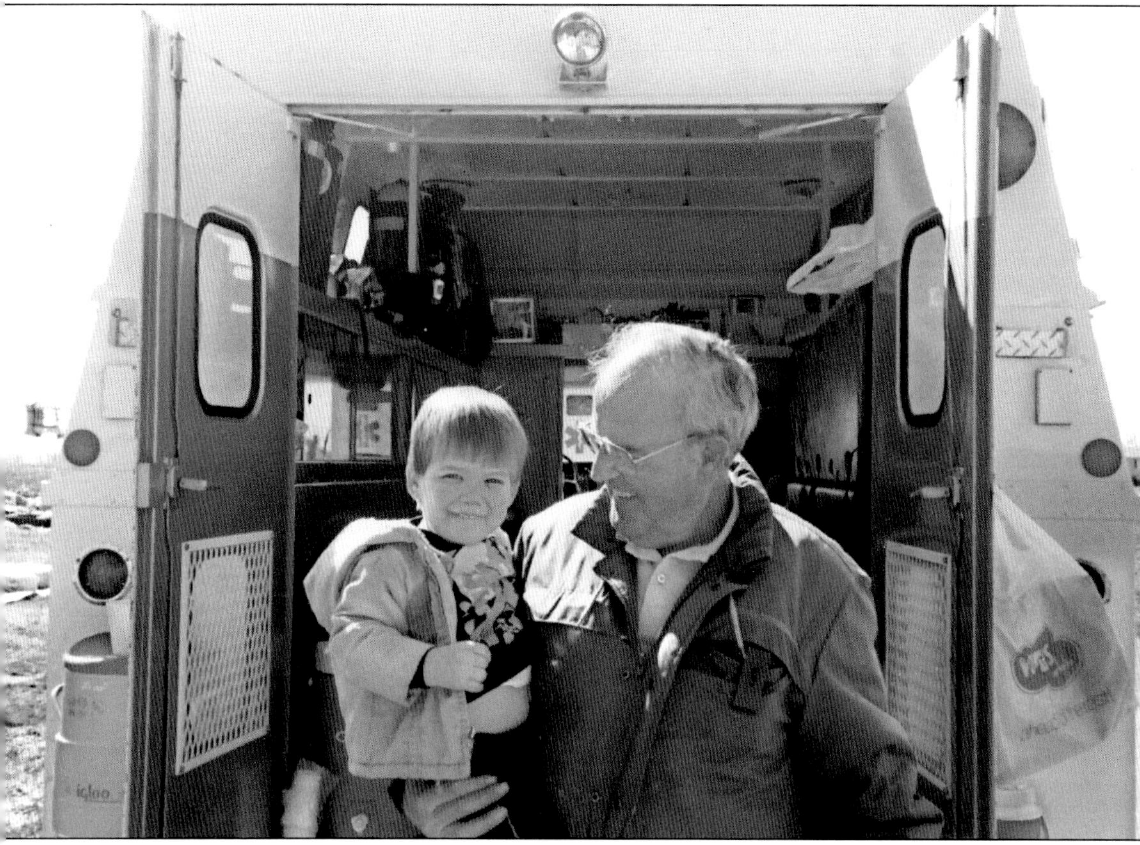

During his lifetime, L. Eugene Johnson responded to hundreds of emergency calls in the Carroll Manor area. He rarely missed responding on a call. He served many years as the rescue chief and the president of the fire company. Rescue Chief Johnson is shown with his grandson Steven in the back of the first Squad 14, a 1981 Ford with a Reading body. This unit is still used as a boat support vehicle. (Courtesy CMFC.)

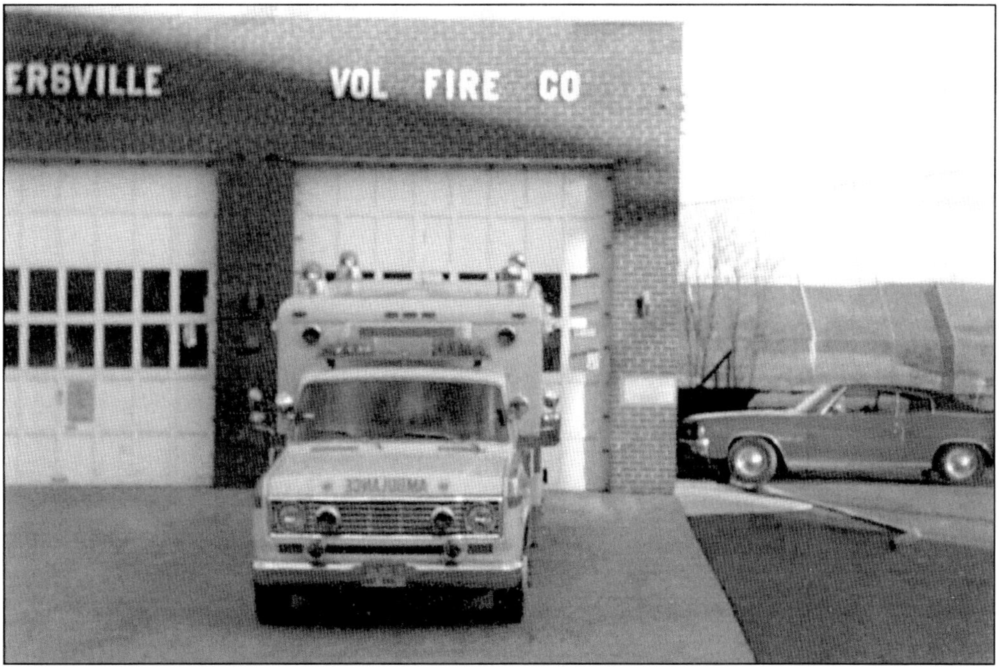

When the Bittle Funeral Home stopped providing emergency ambulance service to Myersville, the fire company answered the call and purchased this 1976 International chassis ambulance. (Courtesy MFC.)

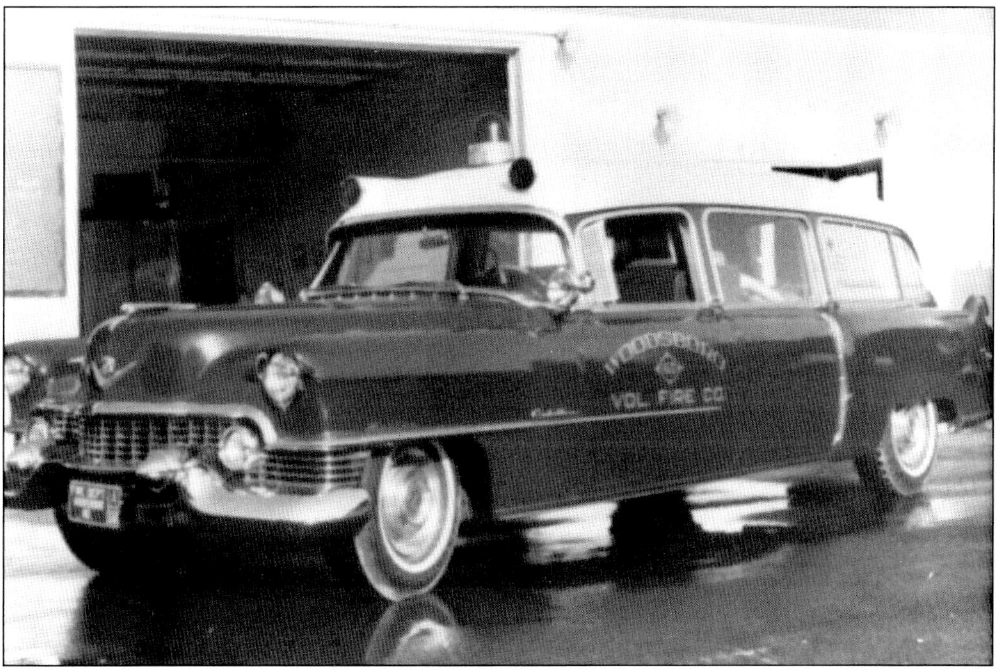

A160 was a 1954 Cadillac ambulance and the first ambulance placed in service by the Woodsboro Volunteer Fire Company. The unit was purchased from the Thurmont American Legion. In the time before Central Alarm was placed in service, most ambulances used 0 rather than 9 in the radio number sequence to designate an ambulance. (Courtesy WBFC.)

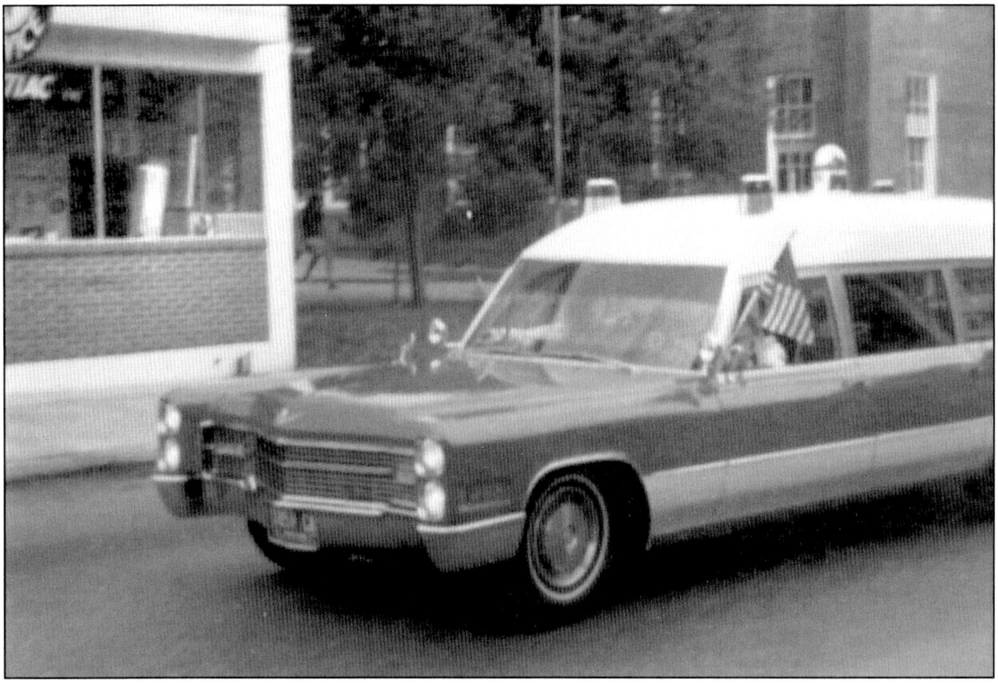

Woodsboro stayed with the Cadillac ambulances when it purchased this 1966 Miller Meteor low-headroom vehicle. Notice the five rotating beacon rays on the roof of the unit rather than the tunnel lights on the Superior Cadillac shown previously. (Courtesy H. Richard Hahn.)

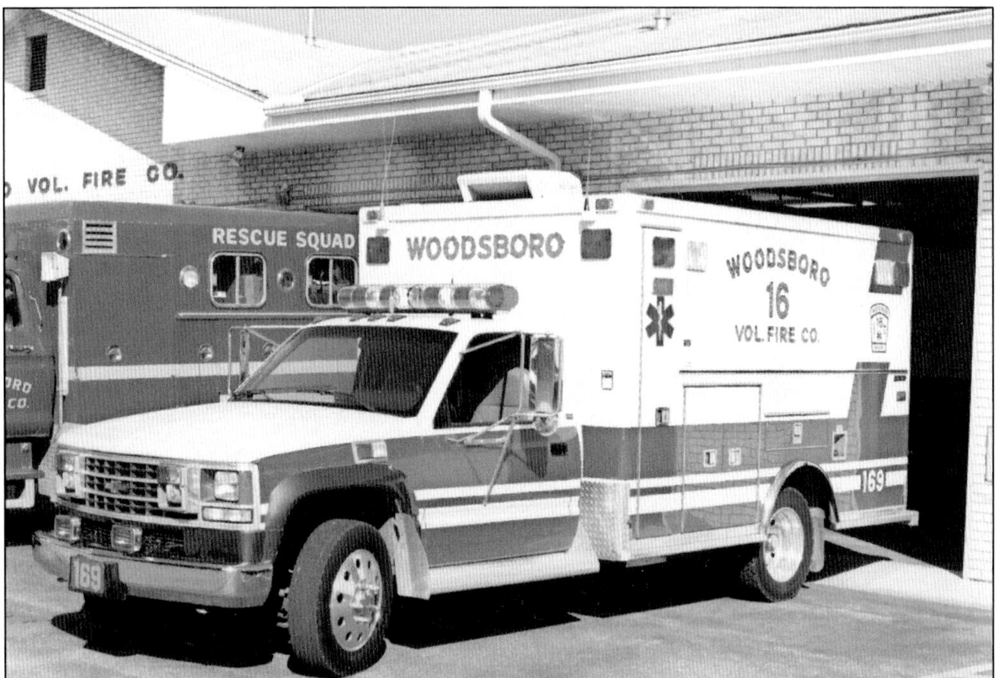

The 1993 Chevrolet Med Tec has recently been retired and is now utilized by the Woodsboro Fire Company as a dive team support unit. (Courtesy Warren Jenkins.)

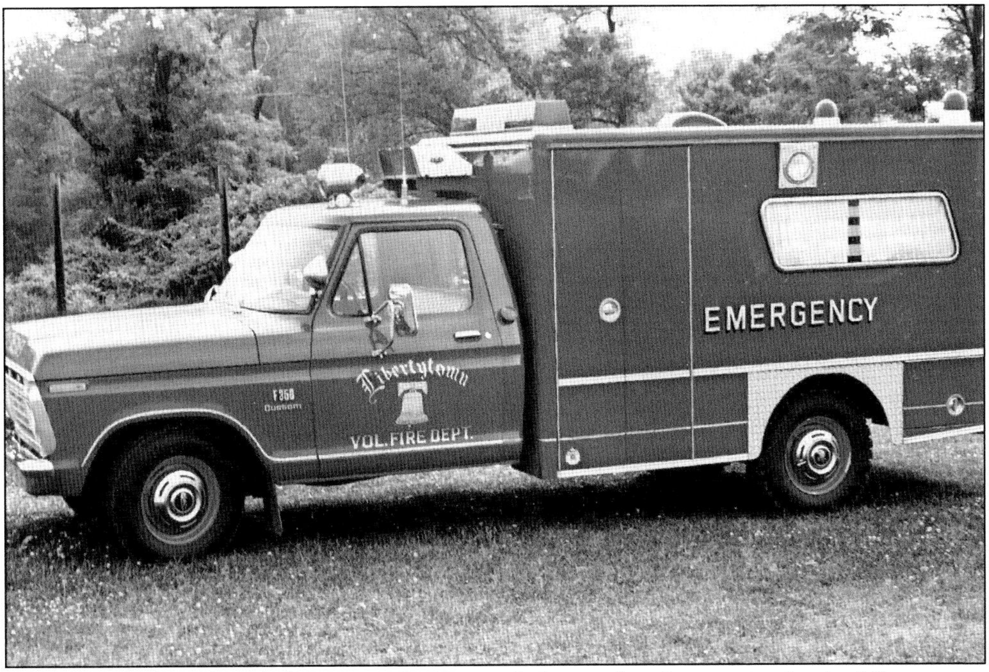

This 1974 Ford Swab ambulance was the first ambulance placed in service by the Libertytown Volunteer Fire Department. This small box unit was purchased used for $3,500 from Delaware. (Courtesy LVFD.)

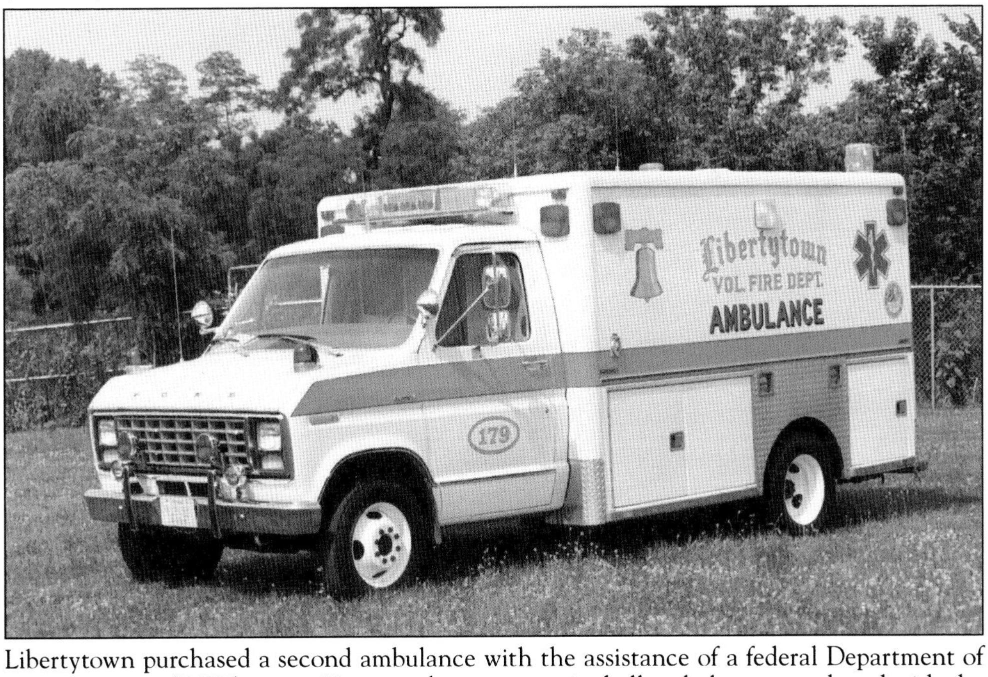

Libertytown purchased a second ambulance with the assistance of a federal Department of Transportation (DOT) grant. However the grant required all ambulances purchased with the DOT money to include orange striping. This 1980 Ford Horton remained in service and was sold to Braddock Heights Volunteer Fire Company when that company began an ambulance service. (Courtesy LVFD.)

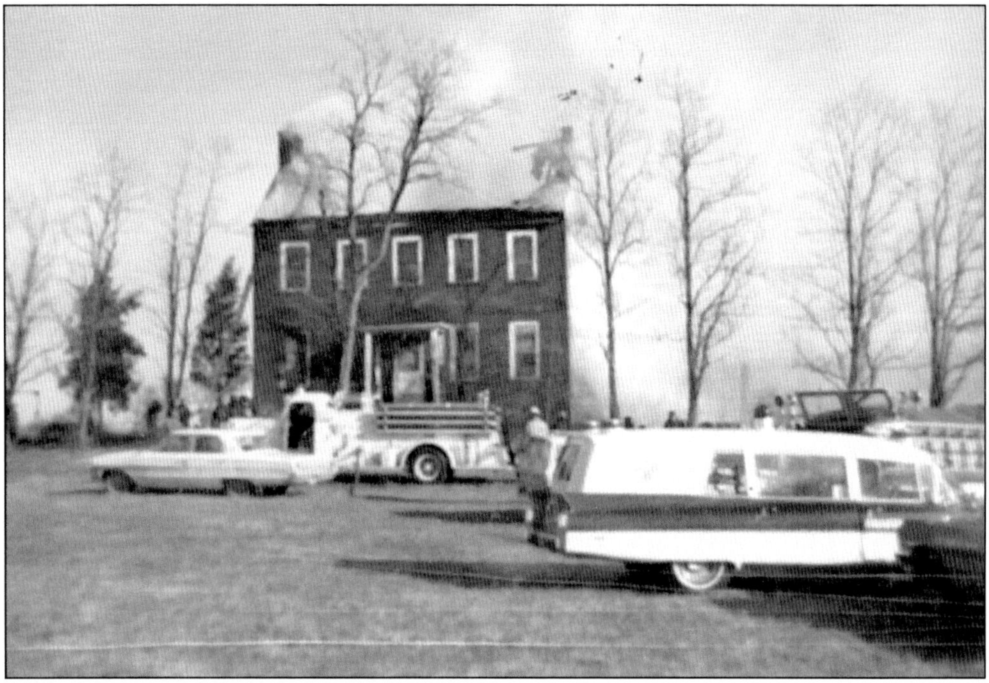

The Graceham Volunteer Fire Company provided ambulance service for several years. One of the Cadillac ambulances owned by Graceham is shown standing by at a training drill with Guardian Hose Company. (Courtesy H. Richard Hahn.)

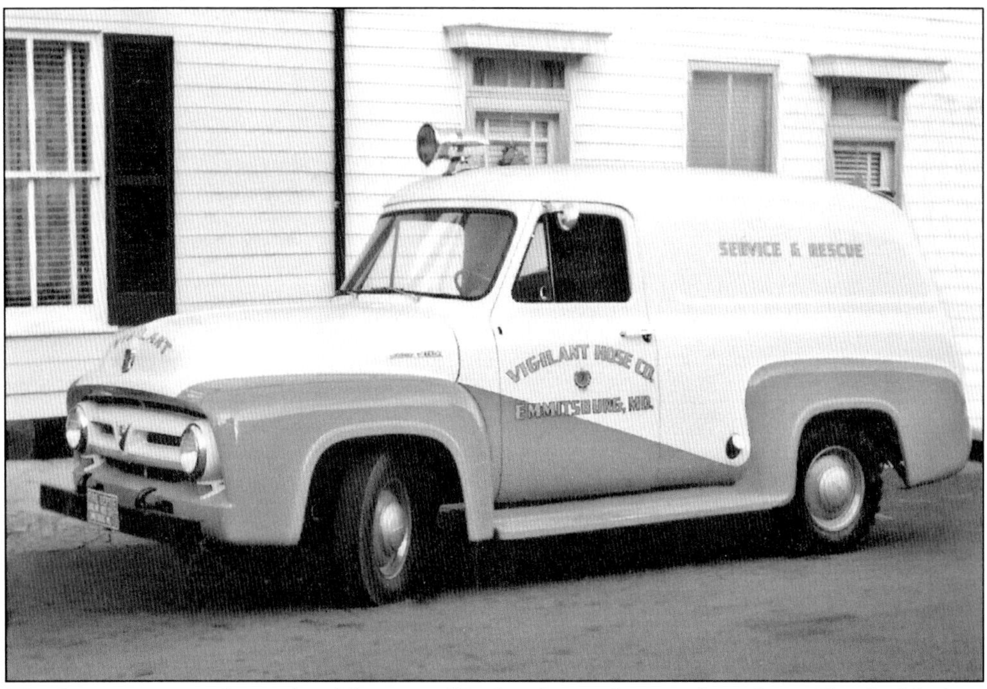

The first rescue squad outside of the city of Frederick was this panel truck used by the Vigilant Hose Company of Emmitsburg. The unit responded on rescue calls as well as fires throughout northern Frederick County. (Courtesy Bob Rosensteel.)

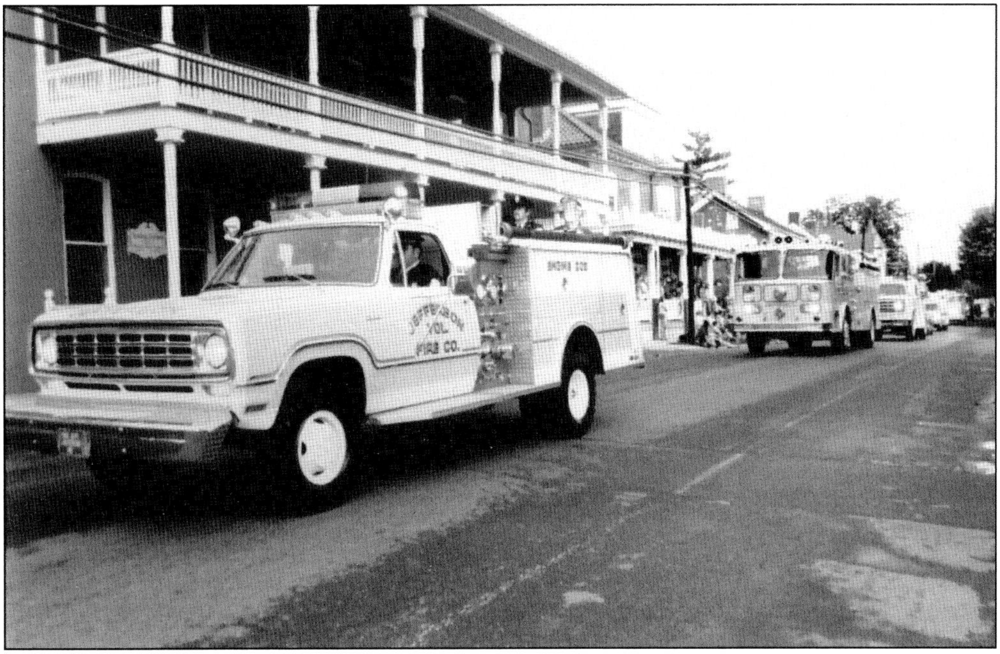

The first non-transport unit in Frederick County equipped to respond to medical emergencies was Special Unit 20. This vehicle was actually Engine 202, a 1976 Dodge four-wheel-drive mini-pumper that would respond to medical calls in the Jefferson area using the designation Special Unit 20. (Courtesy Dave Snyder.)

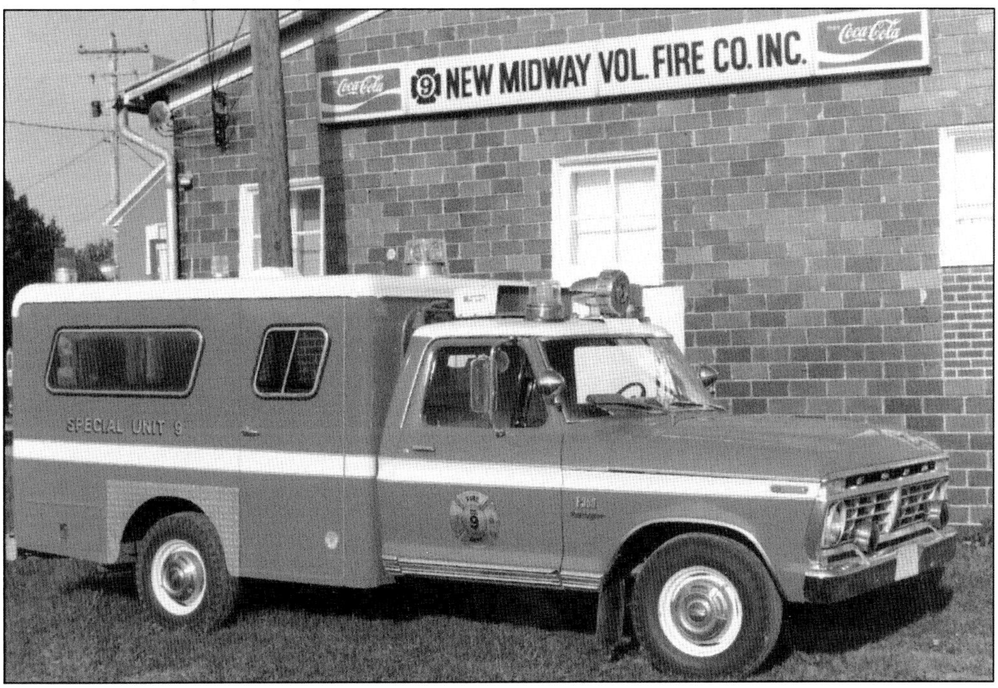

After Jefferson, many fire companies placed special units in service to provide faster emergency medical service. New Midway placed Special Unit 9 in service after they purchased the former ambulance from Graceham. (Courtesy NMFC.)

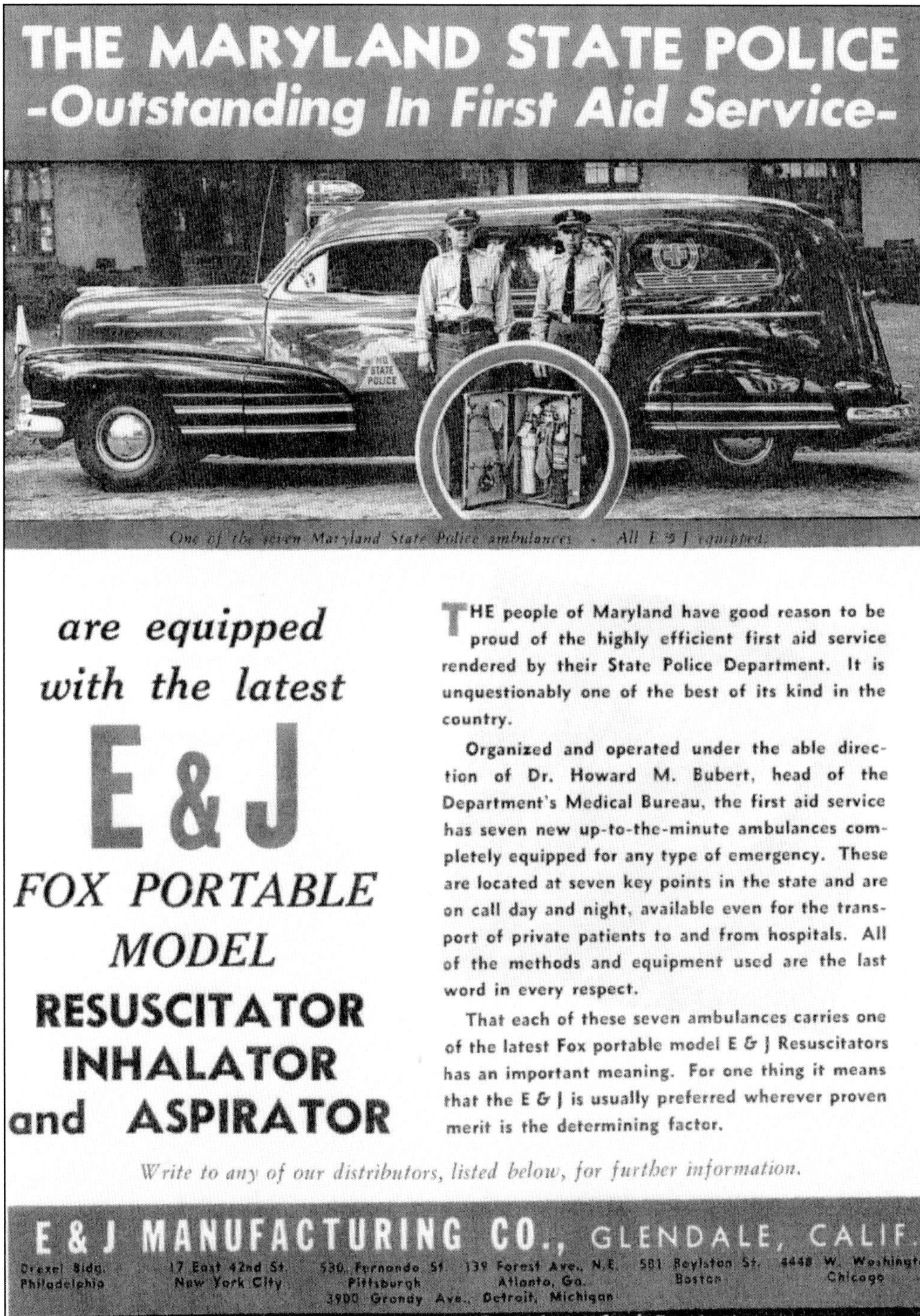

In addition to funeral homes, the Maryland State Police (MSP) maintained ambulance service from seven police barracks well into the 1960s. This image shows an advertisement using an MSP ambulance in the background. The state police had at least one Chrysler ambulance station at the Frederick barrack. (Courtesy JFC.)

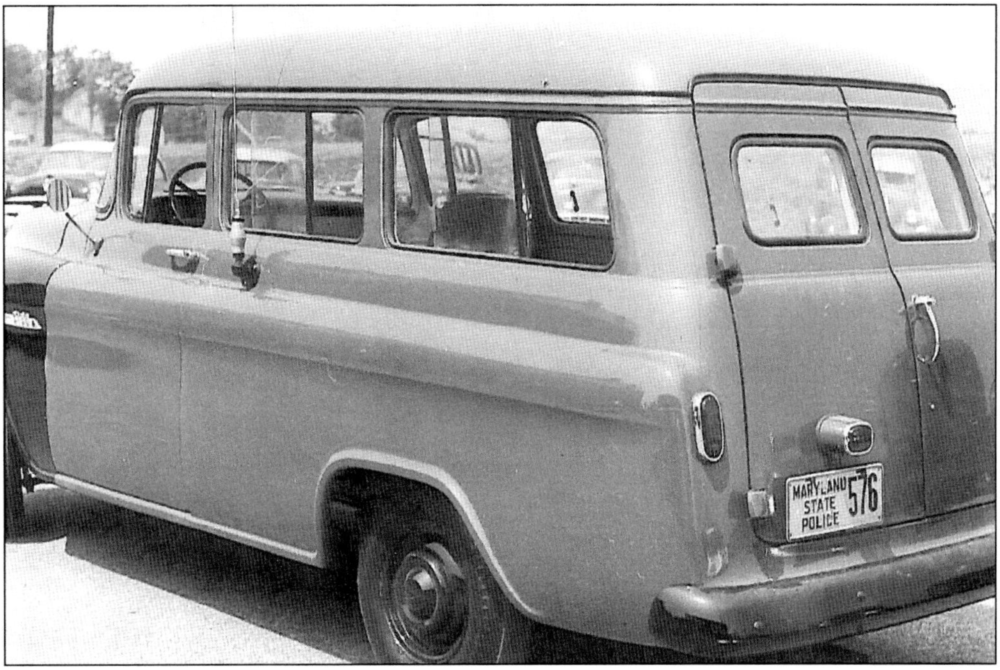

These two pictures show the last MSP ambulances after they were retired from medical service. The Chevrolet Suburbans were the last MSP ambulances to respond from the Frederick barrack. Retired state troopers tell of being called off patrol to go to the barrack to respond to an auto accident with the ambulance. Rarely was there enough manpower to have an extra attendant, so the patients were frequently transported with only a driver in the front. (Courtesy MSP.)

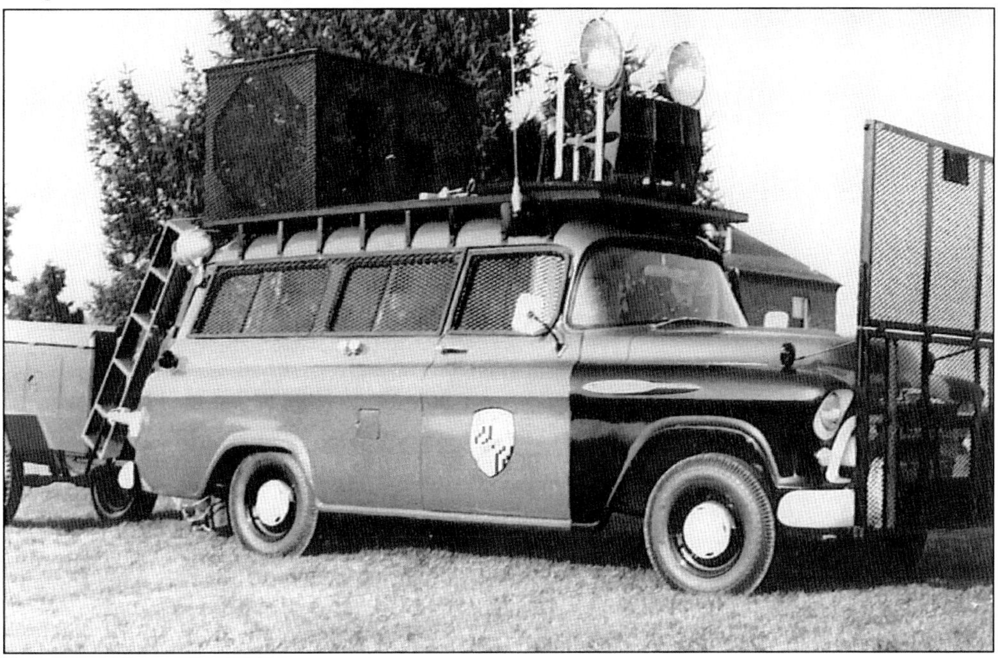

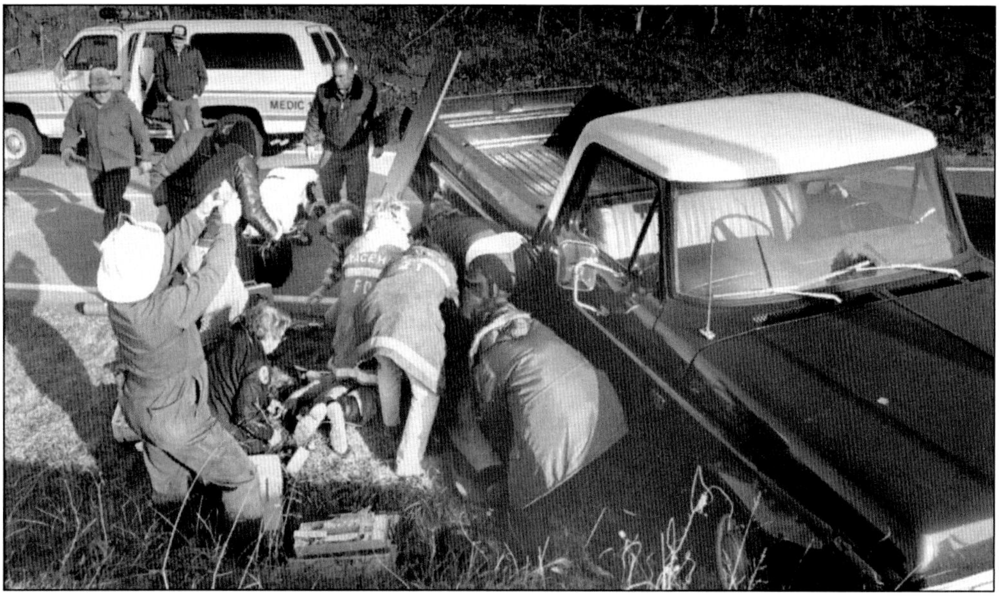

In 1981, Frederick County Advanced Life Support began operation providing volunteer paramedics to respond on critical calls. In the background is Medic 1, a 1981 Suburban and one of the original medic units. Here Medic 1 assists on a serious auto accident. (Courtesy Susie Kyle.)

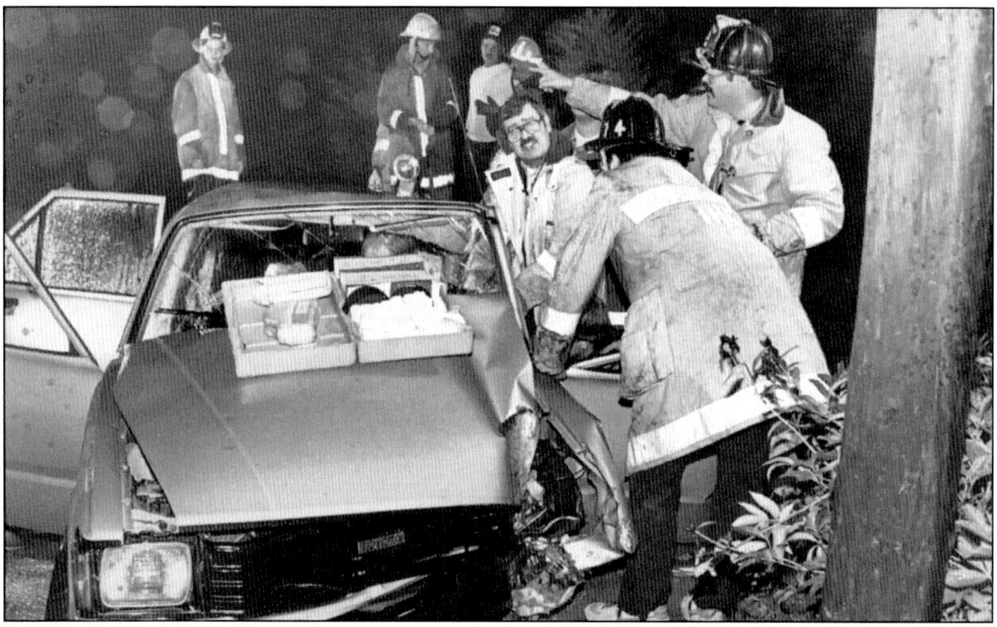

This final photograph shows the dedication and determination of the men and women of the fire and rescue service of Frederick County. Ernie Hughes (left) and Terry Keefer work closely with members from the Independent Hose Company and United Fire Company to extricate and treat seriously injured victims. Mr. Hughes, his wife Margaret, and both children all were active volunteers. Currently, his daughter Edie is a career lieutenant with the Frederick County Division of Fire/Rescue Services, and his son Daniel is a career captain with the Prince George's Fire and Rescue. This is another example of why the fire and rescue service is such a proud family tradition. (Courtesy Susie Kyle.)

Five

People, Places, and Things

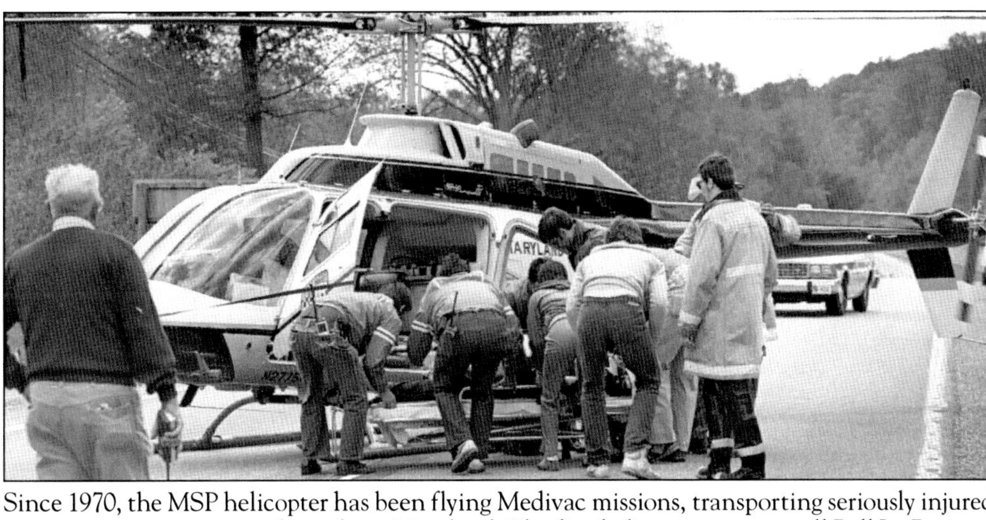

Since 1970, the MSP helicopter has been flying Medivac missions, transporting seriously injured victims to trauma centers throughout Maryland. The first helicopters were small Bell Jet Rangers mainly flown by Vietnam veteran state police troopers. Here a victim from a serious accident on Route 15 near Thurmont is loaded for a transport to the Shock Trauma Center in Baltimore. (Courtesy Susie Kyle.)

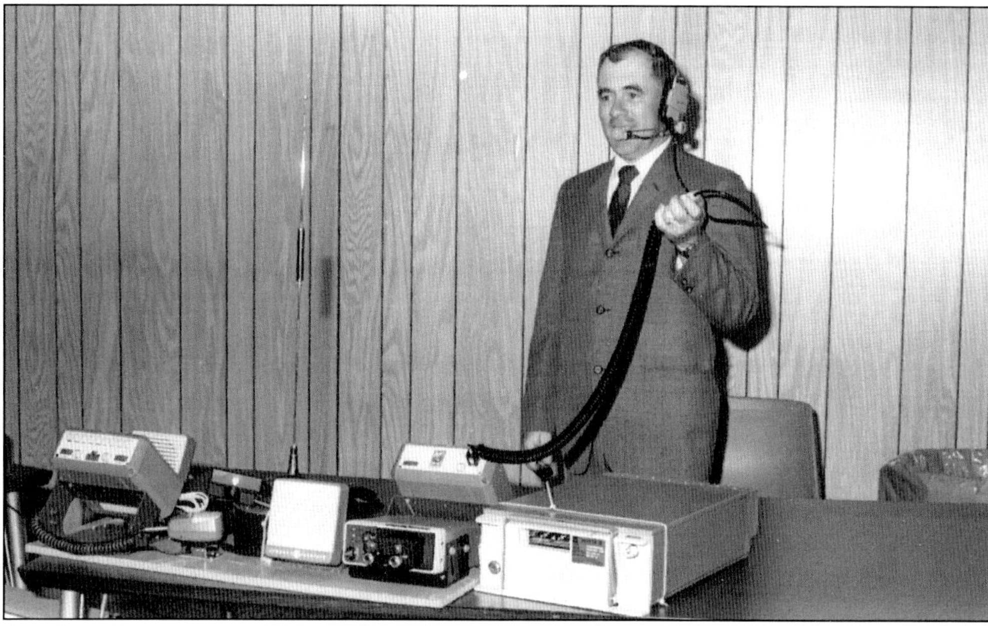

Another advancement in Maryland was the implementation of a statewide radio system allowing any ambulance in the state to communicate with any hospital. This massive system was coordinated by the Maryland Institute for Emergency Medical Services (MIEMS). Shown is Mike Smith, the first regional coordinator for Region II, Frederick and Washington Counties, in this 1974 picture demonstrating the new radios.

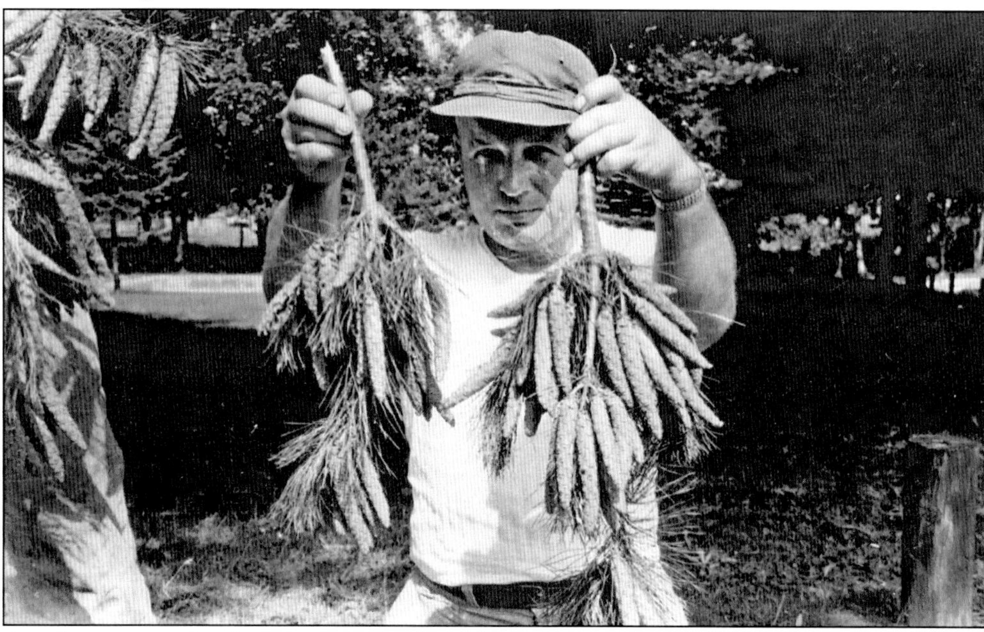

Another agency that has worked closely with the fire service is the Department of Natural Resources, previously called the Forestry Department. In this picture from August 1956, Herman Toms, a local ranger and member of the Braddock Heights Volunteer Fire Department, shows white pine seedlings that are about to be planted in the Frederick Watershed. (Courtesy Frederick Fire Museum, donated by Marie Kendall.)

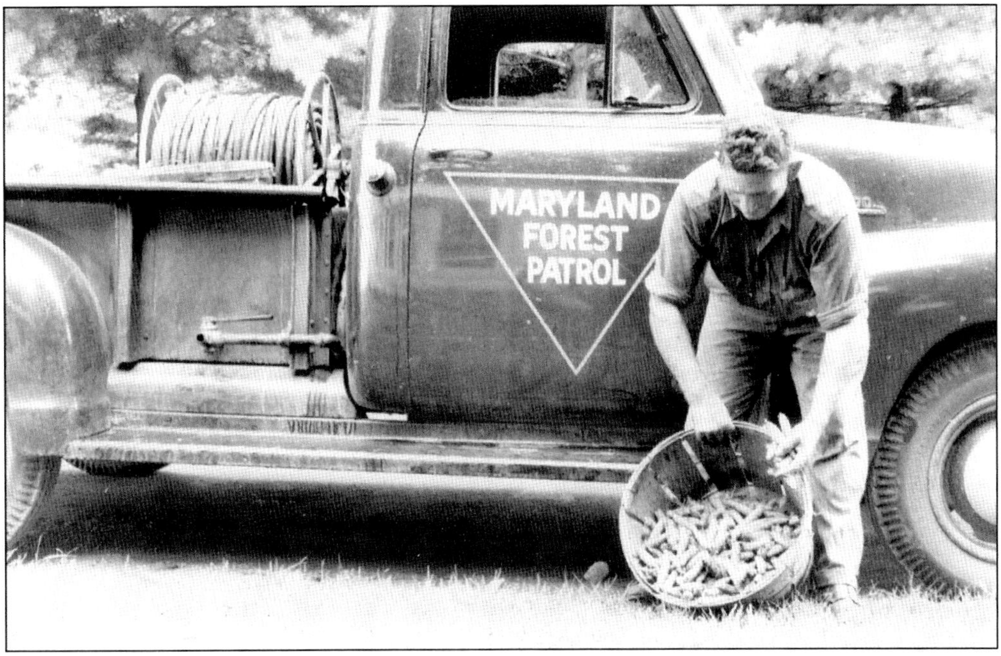

Ranger Donald Kendall is shown next to his Maryland Forest Patrol firefighting vehicle in this 1956 picture. He and Herman Toms worked together and helped local firefighters fight woods and mountain fires well into the 1970s. (Courtesy Frederick Fire Museum, donated by Marie Kendall.)

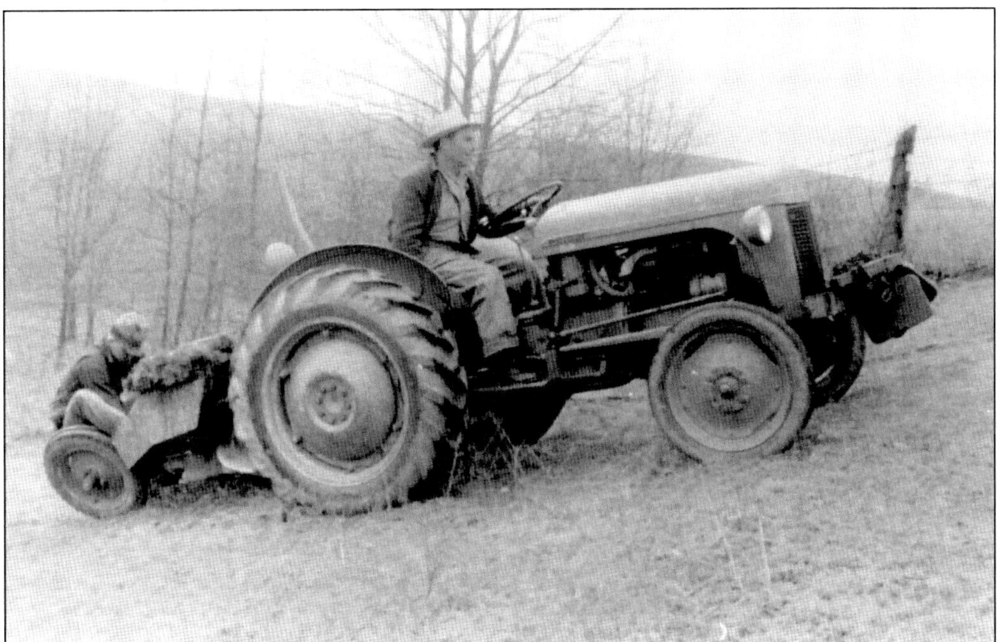

In 1956, additional trees were planted to help ensure the stability of the area of the Linganore watershed that protects the water supply for the city of Frederick. Forest rangers are shown using a tractor to help plant the seedlings. (Courtesy Frederick Fire Museum, donated by Marie Kendall.)

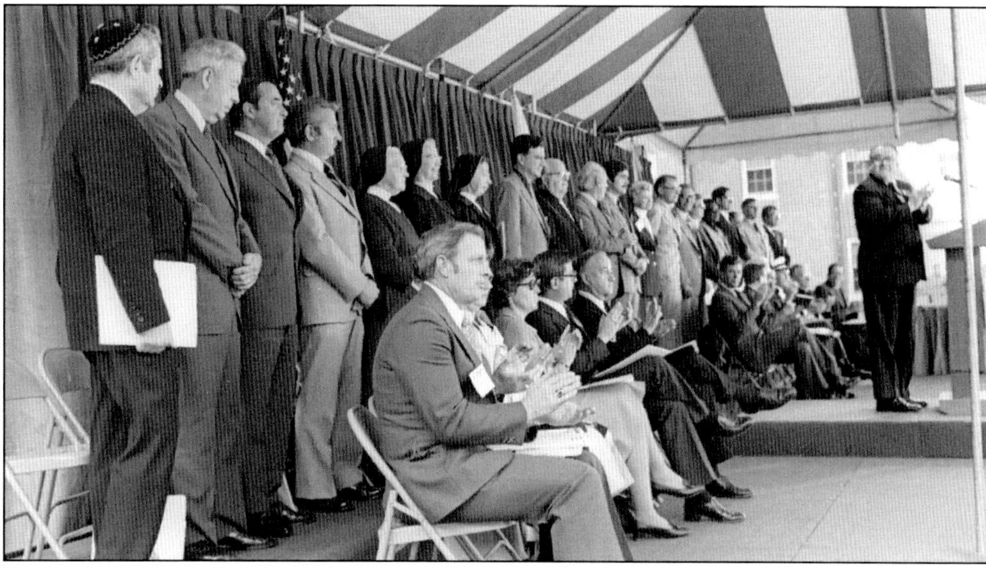

Frederick County is home to the National Fire Academy, which is located on the former St. Joseph's College campus in Emmitsburg. Dignitaries that are identifiable in this ceremony opening the facility are Gov. Harry Hughes, Sen. Charles "Mac" Mathias, and Congresswoman Beverly Byron. (Courtesy Susie Kyle.)

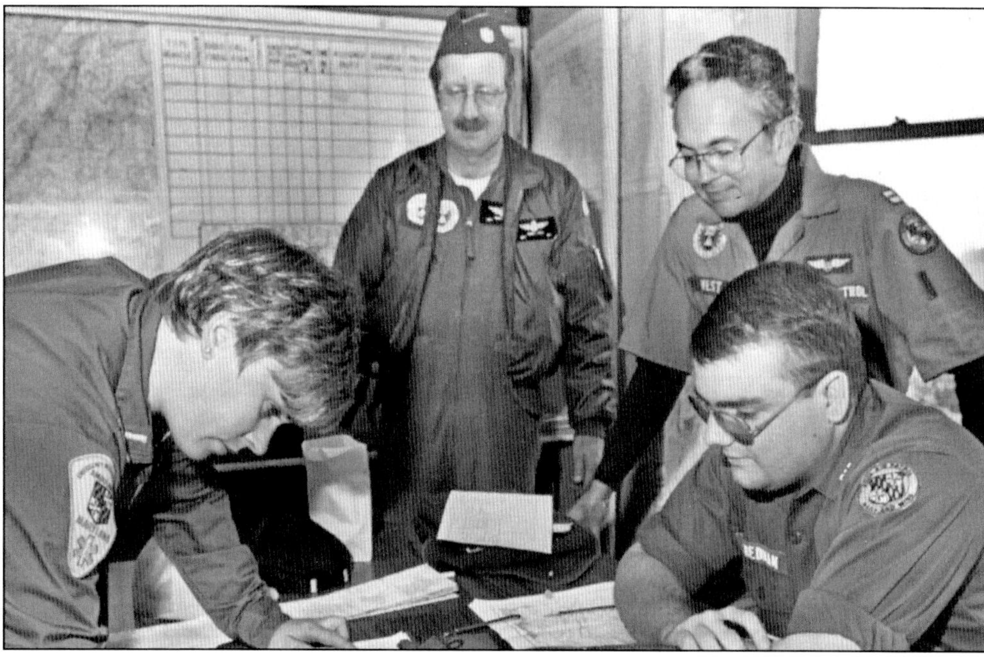

The Civil Air Patrol (CAP) has played a vital role in assisting local authorities to search for lost aircraft for over 50 years. The CAP and many public safety agencies were called to find one of their own when the Maryland State Police helicopter was reported missing in January 1986. As they plot the county map, an extensive search was conducted in and around Frederick County. The remains of the Frederick-based helicopter were located in a park at the end of I-70 in Baltimore City. Unfortunately, two dedicated Maryland state troopers died in this tragic aircraft accident. (Courtesy Susie Kyle.)

Long before Homeland Security, Civil Defense provided information to the general public to prepare for a major emergency. This 1956 instruction card was a common site in most firehouses and public buildings.

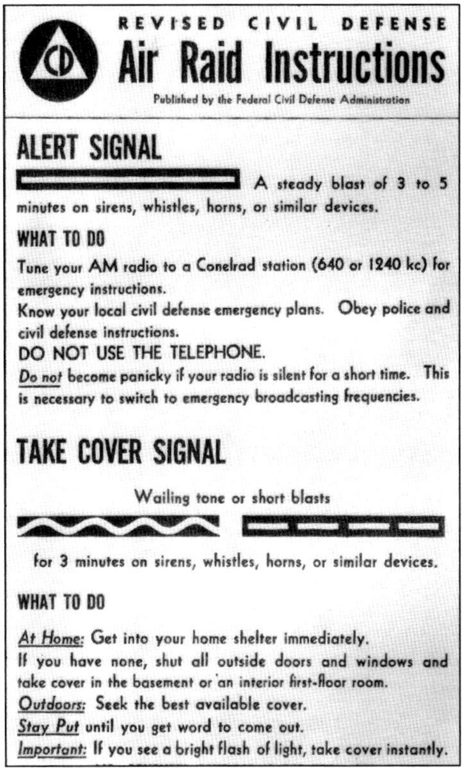

REVISED CIVIL DEFENSE

Air Raid Instructions

Published by the Federal Civil Defense Administration

ALERT SIGNAL

A steady blast of 3 to 5 minutes on sirens, whistles, horns, or similar devices.

WHAT TO DO

Tune your AM radio to a Conelrad station (640 or 1240 kc) for emergency instructions.

Know your local civil defense emergency plans. Obey police and civil defense instructions.

DO NOT USE THE TELEPHONE.

Do not become panicky if your radio is silent for a short time. This is necessary to switch to emergency broadcasting frequencies.

TAKE COVER SIGNAL

Wailing tone or short blasts

for 3 minutes on sirens, whistles, horns, or similar devices.

WHAT TO DO

At Home: Get into your home shelter immediately.

If you have none, shut all outside doors and windows and take cover in the basement or an interior first-floor room.

Outdoors: Seek the best available cover.

Stay Put until you get word to come out.

Important: If you see a bright flash of light, take cover instantly.

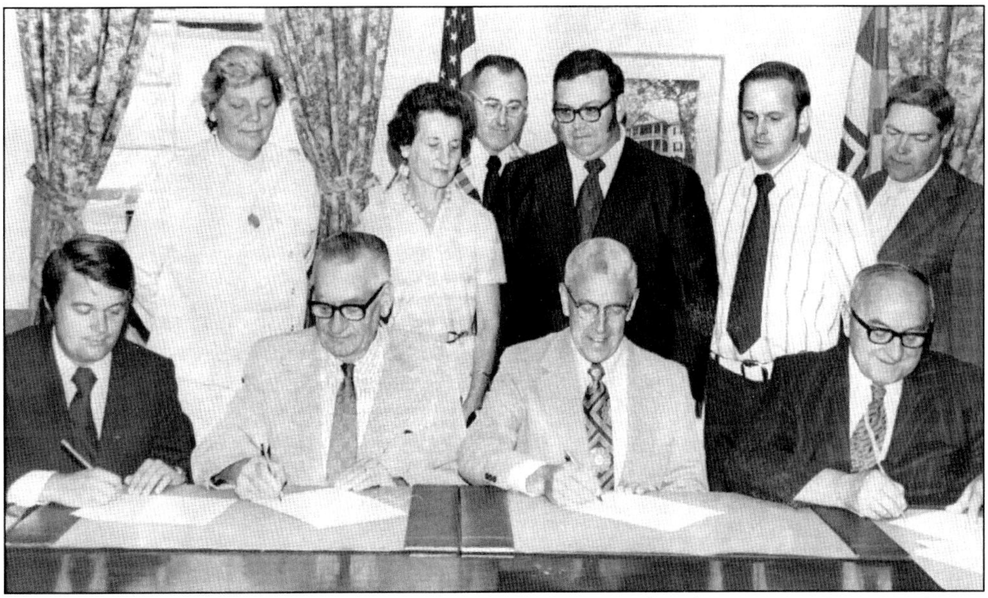

From left to right in this picture for EMS Week in 1974 for Frederick County and Frederick city, Mayor Ron Young and county commissioners John Derr, Donald Lewis, and Lawrence Dorsey sign a proclamation. In the second row, from left to right, Agnes Kemmerer, EMT instructor and nurse; Mary Hargis, Heart Association; Dr. Robert Roberts, EMT instructor; and Frederick County Ambulance Association members John Knipple, Dave Late, and Bernard Williams look on. (Courtesy Susie Kyle.)

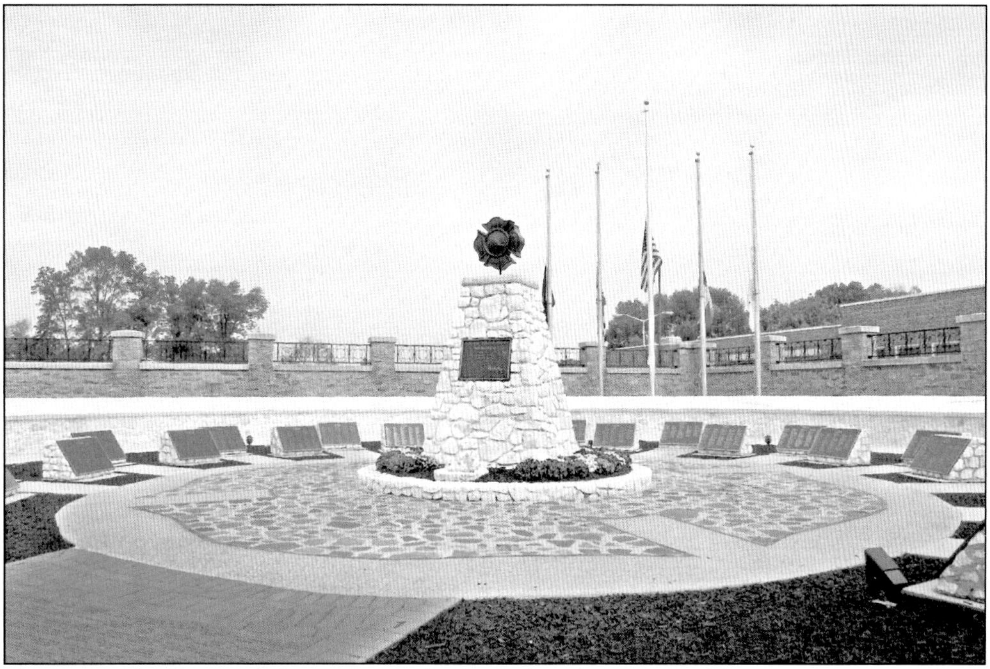

The National Fallen Firefighters Memorial is located on the campus of the National Fire Academy in Emmitsburg. Each year, the families of firefighters that have died in the line of duty gather in Emmitsburg for a memorial service organized by the National Fallen Firefighters Foundation. (Courtesy National Fallen Firefighters Foundation.)

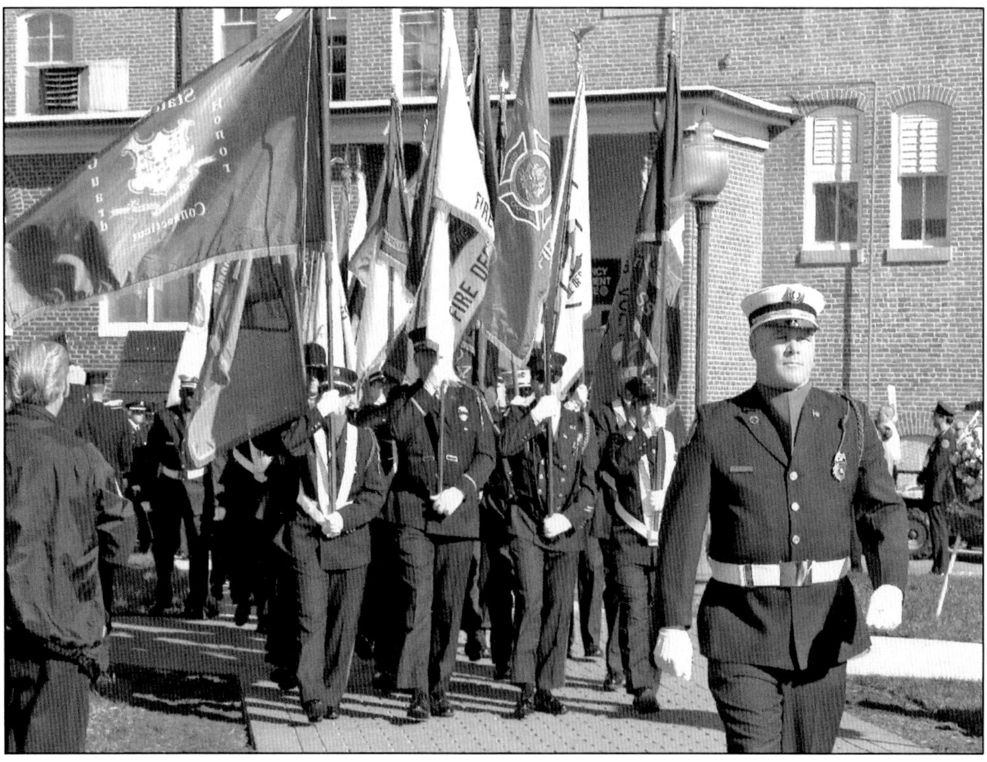

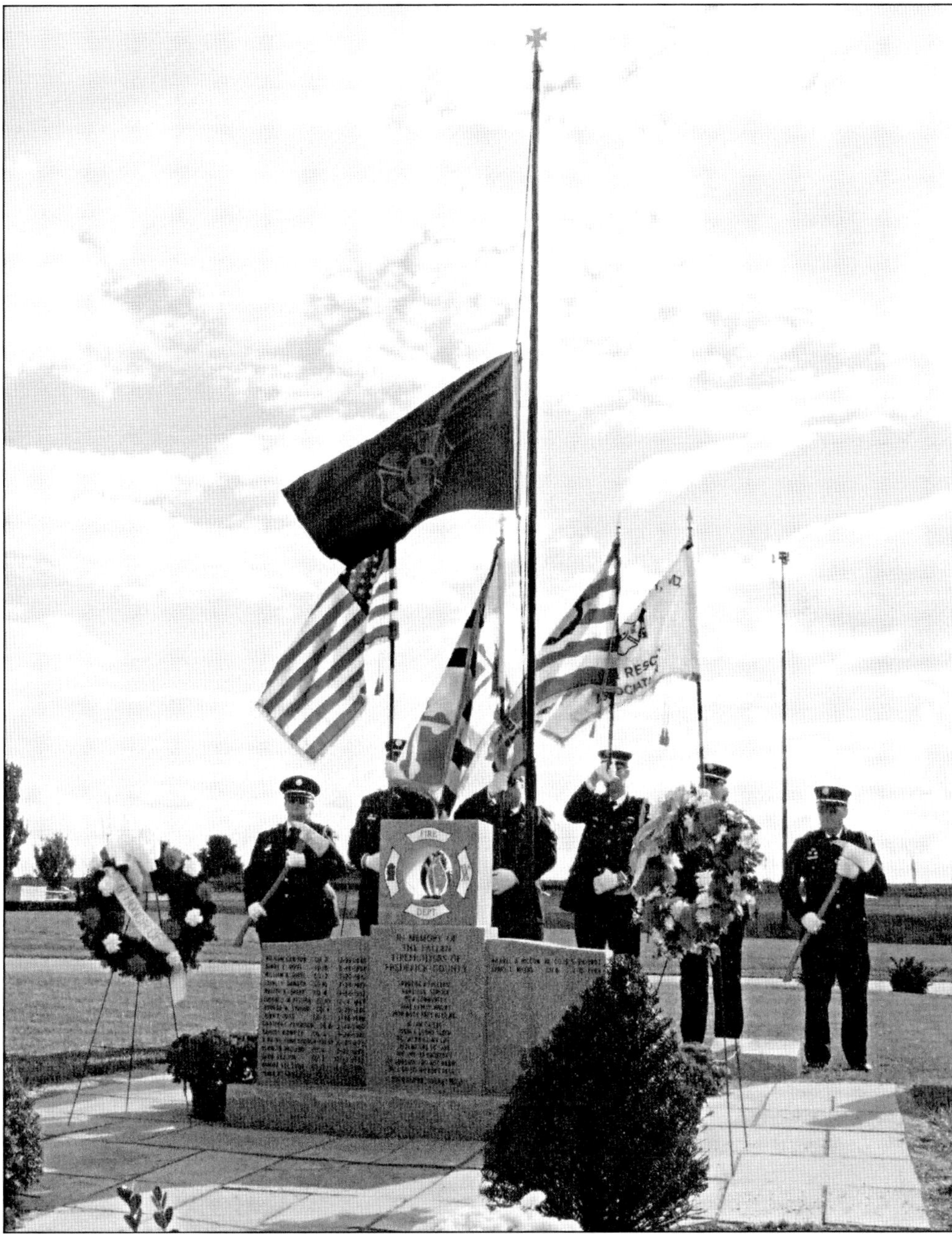

A memorial to all Frederick County firefighters that have died in the line of duty is located at Mount Olivet Cemetery in Frederick. The Honor Guard of the Frederick County Volunteer Fire and Rescue Association is shown at the annual memorial service honoring all county fire and rescue personnel that died in the previous year. (Courtesy David Snyder.)

PROGRAM

Thurs., August 13th.; Fri., August 14th.; Sat., August 15th, 1936

Thursday, August 13th.

Headquarters and Place of Meeting—Fire Engine House

6:00 P. M.—Registration and Reception of Delegates

8:00 P. M.—Convention to Order

Singing America

PRAYER ....Chaplain, William S. Wachtel

ADDRESS OF WELCOME................................Mayor Michael J. Thompson

RESPONSE......................F. Lester Smith, Past President, Frederick, Md.

Current Business, Nomination and Election of Officers; Convention Town for 1937; Next Quarterly Meeting Place.

Luncheon

FRIDAY, AUGUST 14th.

5:30 P. M.—Parade of Fire Companies, Bands and Civic Organizations. Parade will form on S. Frederick St., opposite Emmitsburg R. R. Station.

Organized in 1935, the second convention of the Frederick County Firemen's Association was hosted by the Vigilant Hose Company of Emmitsburg. This page was taken from an original program book printed for the Emmitsburg convention, held August 13, 14, and 15, 1936. (Courtesy Leo Boyle.)

When first placed in service, Frederick County Central Alarm was located on Montevue Lane in the former Frederick Emergency Hospital on the Montevue County Home property. This photograph, taken in the late 1970s, shows Randy Waesche, the former bureau chief of the Emergency Communications Center, working alone at the Motorola console. In the early days, many shifts were manned by just one dispatcher on duty. (Courtesy Dave Snyder.)

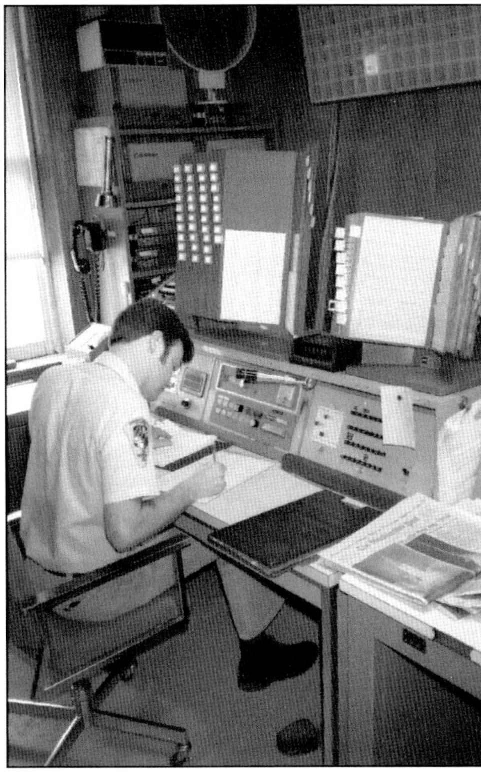

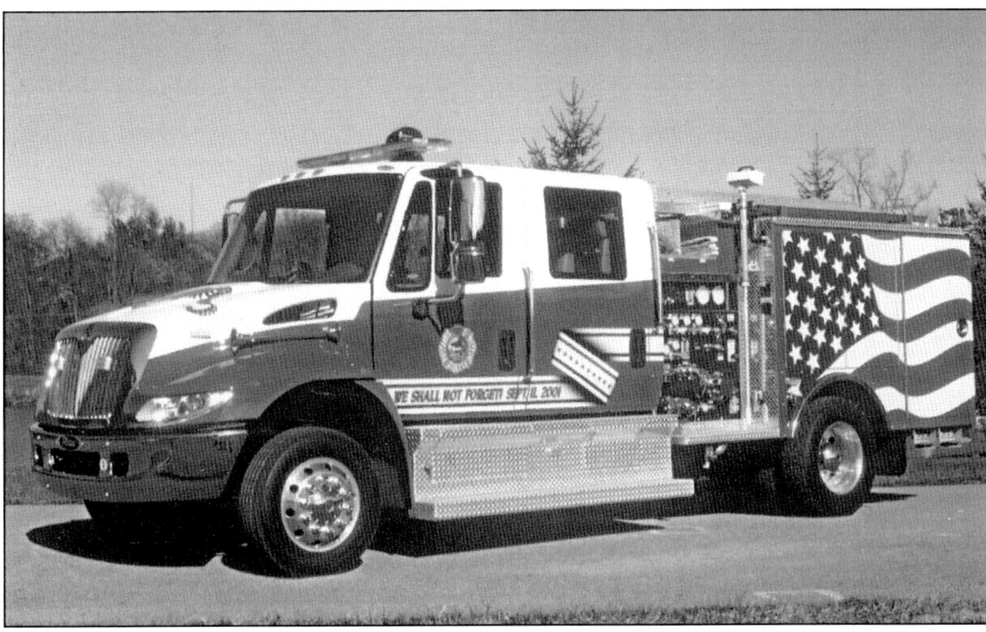

Since Franklin Delano Roosevelt was president, the Catoctin Mountains of Frederick County have been a refuge for the commander in chief at Camp David. Shown at the training center in Frederick, this patriotic 2002 International Pierce engine protects the retreat and surrounding areas. The engine has a tribute to September 11, 2001, along the cab doors. (Courtesy Joel C. Woods.)

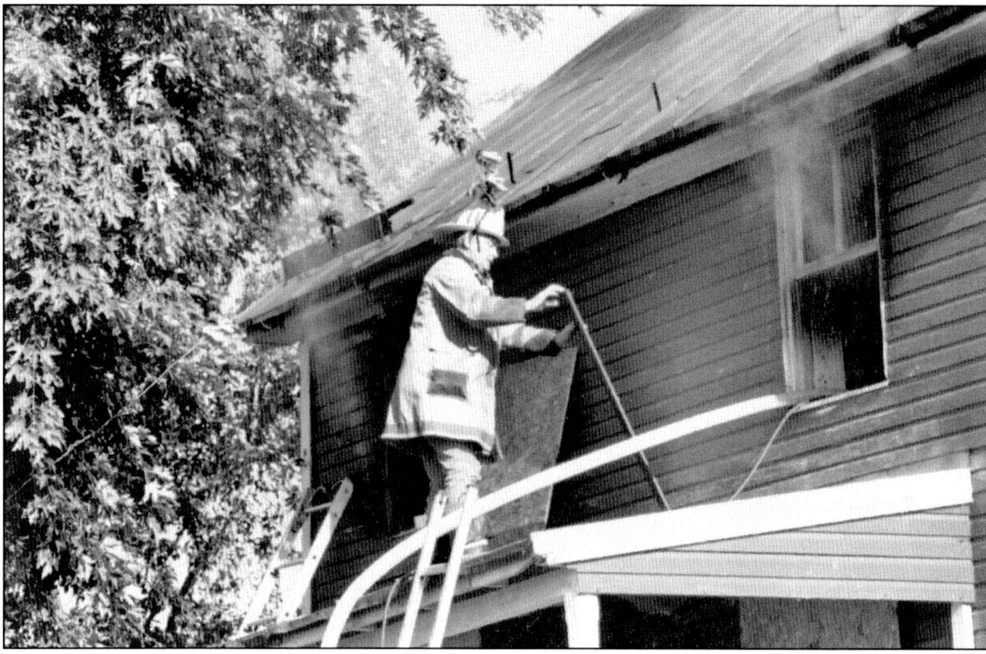

Training has always been paramount for the members of the fire and rescue service in Frederick County. Chief Bob Campbell of the Walkersville Volunteer Fire Company prepares a house for a training drill for members of his department in the summer of 1991. (Courtesy WVFC.)

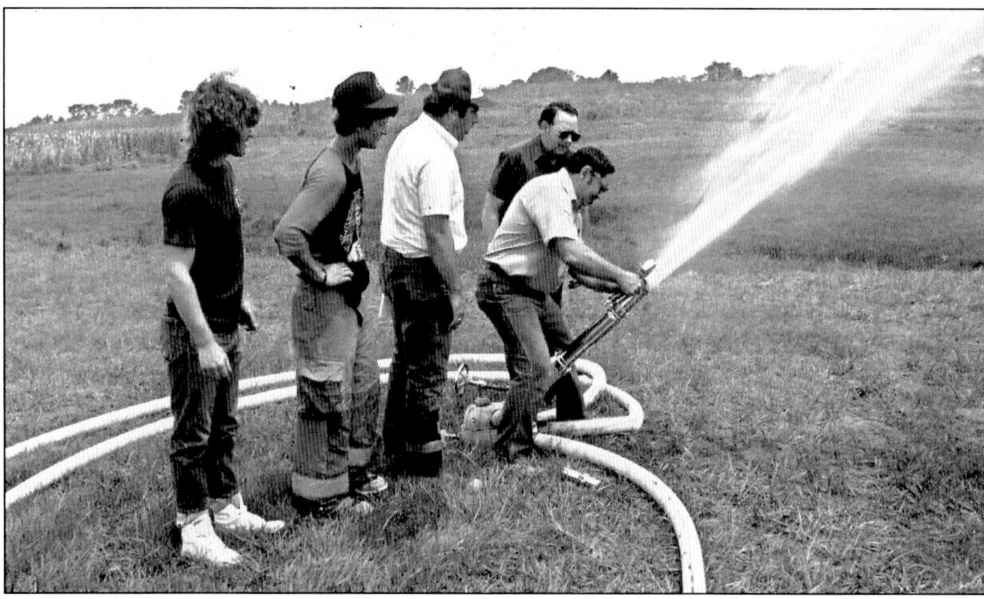

A longtime instructor for the Maryland Fire Rescue Institute and volunteer at the New Midway Volunteer Fire Company, Bob Renner (at the nozzle) uses a pitot gauge to measure the amount of water flowing from a pumper during a pump test at the property of Charlie Smith (standing behind with hat), a former member of the Maryland House of Delegates and past president of the Brunswick Volunteer Fire Company. Merhl Remsberg, next to Mr. Renner, was a longtime chief of the Jefferson Volunteer Fire Company. The two firefighters on the far left are unidentified. (Courtesy Dave Snyder.)

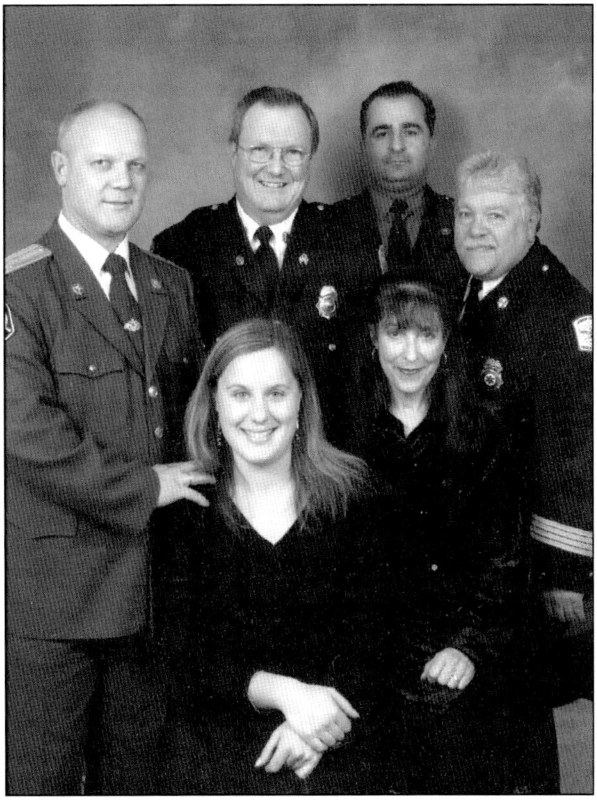

Dr. Marshal Botkin, an assistant chief of the Junior Fire Company No. 2 and a professor at Frederick Community College, has developed a unique international relationship with members of the fire service in Russia. He has traveled there and participated with the Russian Fire Service. In January 2005, three members of the Russian Fire Service visited the United States and attended the annual banquet of the Juniors. Assistant Chief Botkin; his wife, firefighter/EMT Sharon Botkin; and the author, Clarence "Chip" Jewell, Frederick County director of Volunteer Fire and Rescue Services, are shown in full dress with Russian fire officers St. Petersburg District chief Vladimer Sukhanov, Engine Company 7 chief Merab Panchevidce, and Maj. Anna Vorbeitchicova of the Russian National Fire Academy. (Courtesy Junior Fire Company.)

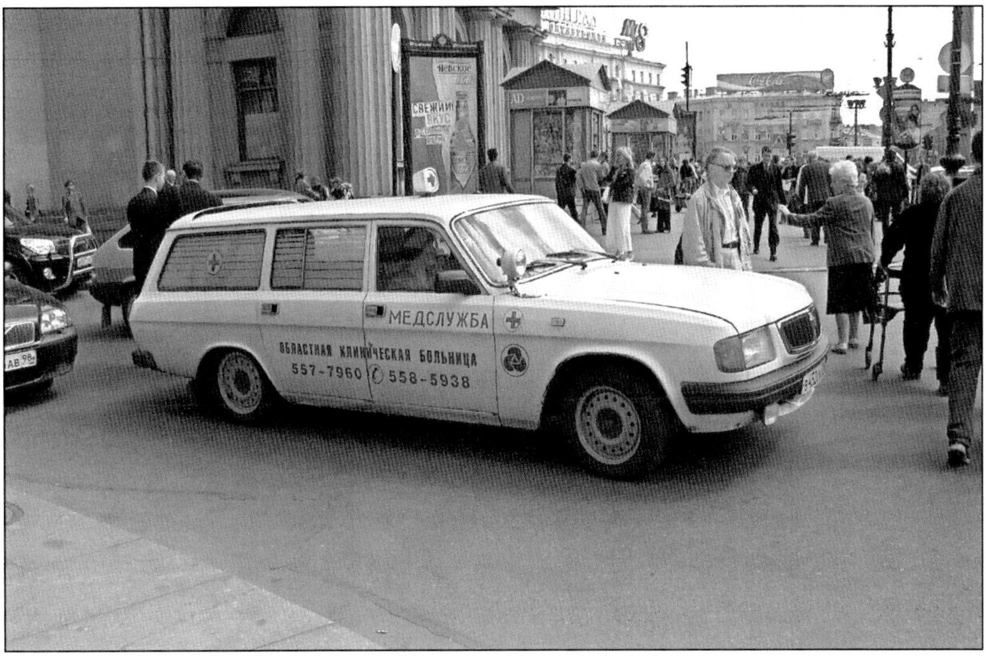

In a picture taken by Assistant Chief Botkin during one of his Russian trips, an ambulance providing medical service in Russia is quite a bit smaller than the American counterparts. (Courtesy Marshal Botkin.)

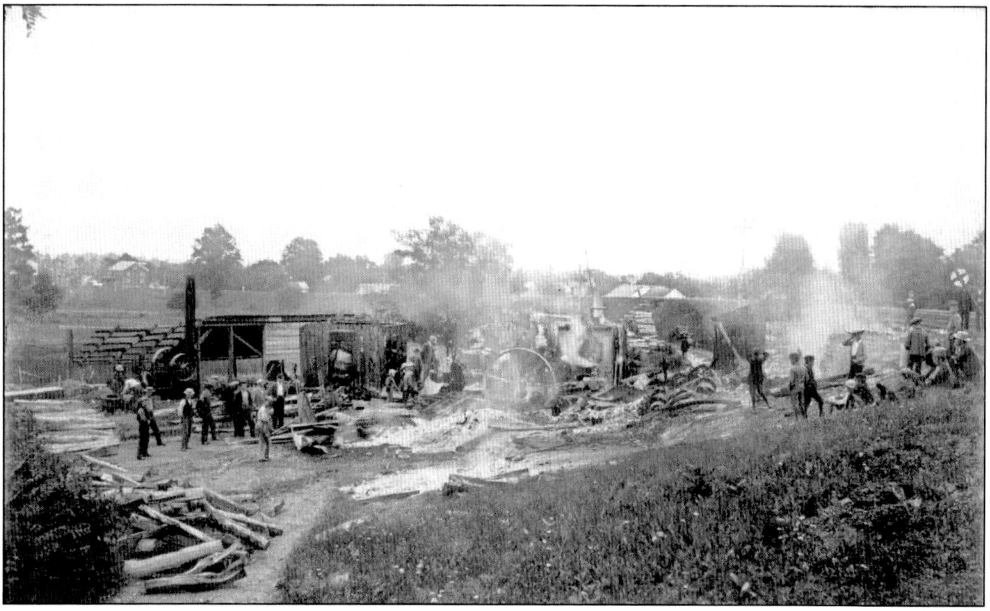

Firefighting has always been a dangerous activity. Since 1840, 18 firefighters in Frederick County have given their lives in the line of duty. Only one death was a result of injuries actually fighting a fire. In 1898, a fire consumed the Excelsior Mill in Thurmont. Nearing the end of the firefighting operations, a large boiler smokestack collapsed, killing Harry C. Root, a member of the Guardian Hose Company. (Courtesy Thurmont Historical Society.)

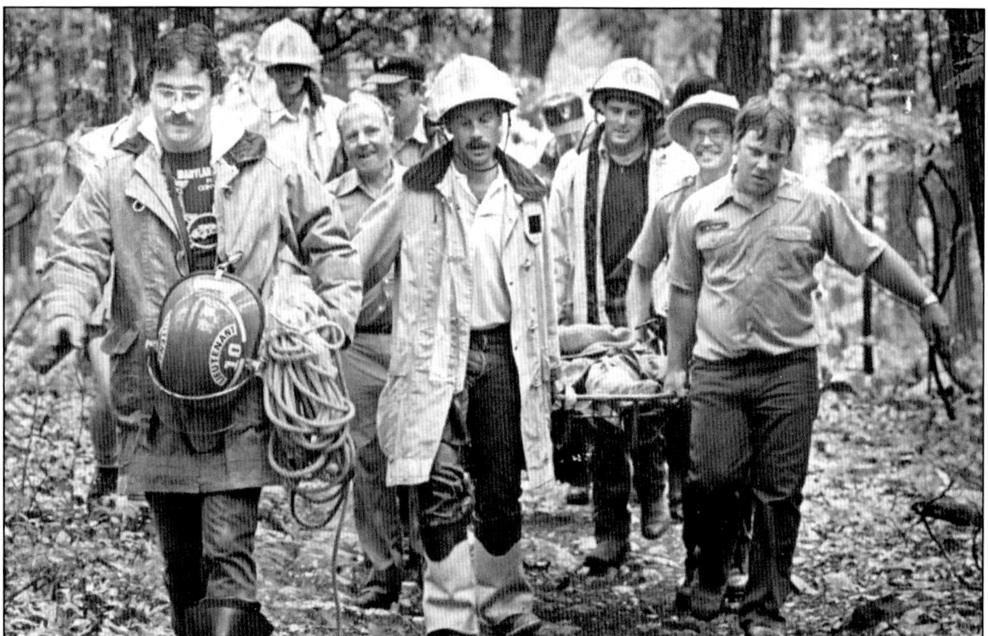

The National Park Service is responsible for the Chesapeake and Ohio Canal National Park and the large Catoctin Mountain Park west of Thurmont. In May 1986, members of the Guardian Hose Company of Thurmont are shown assisting the National Park rangers in moving a victim from the mountainous area to an awaiting ambulance for transport. (Courtesy Susie Kyle.)

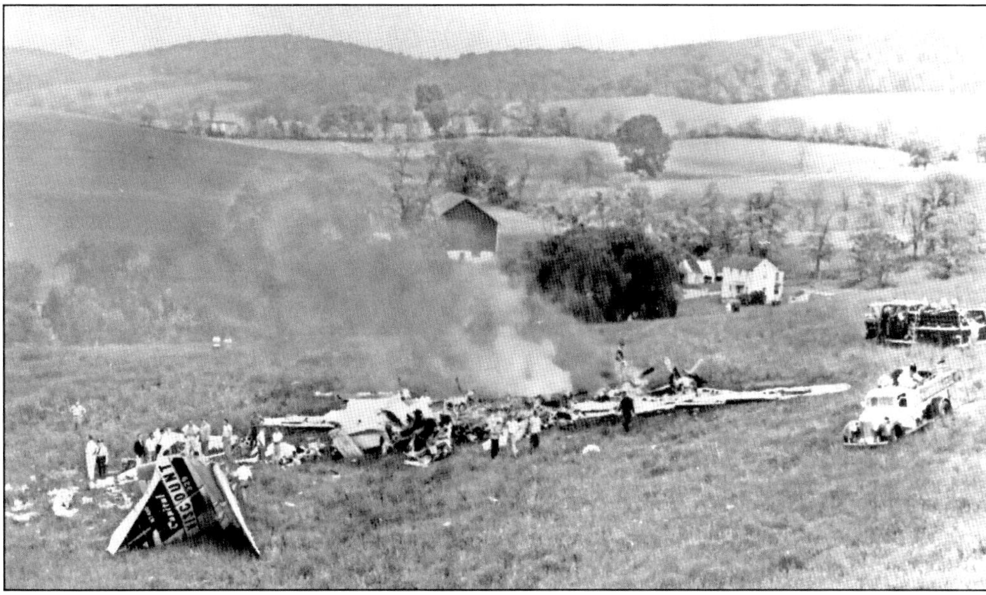

On May 20, 1958, an air disaster struck Frederick County as a Capital Airlines Viscount was struck in midair by a Maryland National Guard jet trainer near Jefferson. The pilot of the National Guard aircraft parachuted to safety, while 12 civilians died as the plane crashed into a farm field. Fire department and rescue units from throughout Frederick County responded. This incident was only the fourth call for the newly formed Brunswick Ambulance Company. Thirty-four additional passengers had just been dropped off in Pittsburgh prior to the crash. (Courtesy CMFC.)

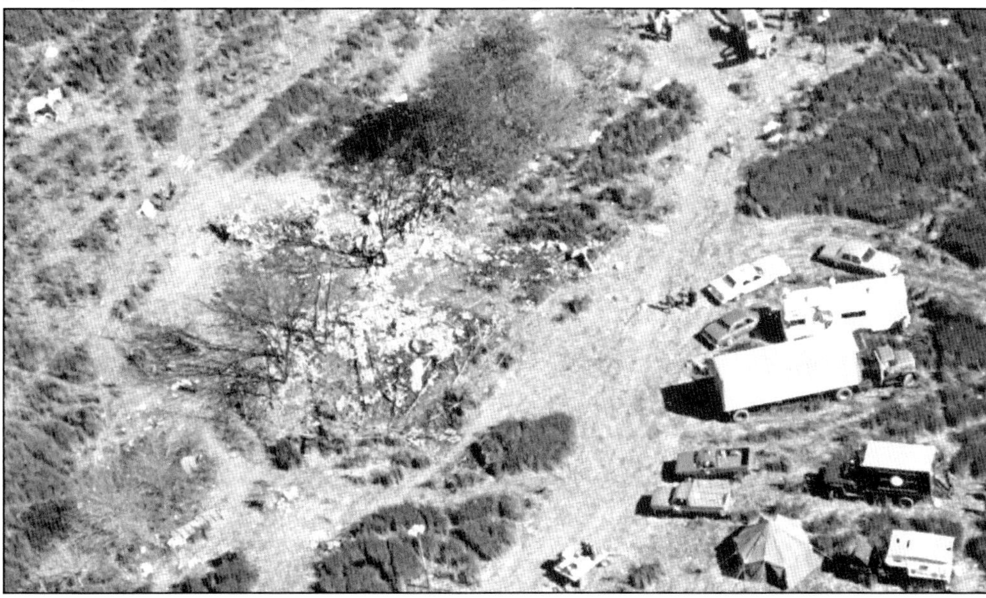

The largest aircraft accident in the history of Frederick County came on May 6, 1981. Military aircraft, flying at an altitude of over 50,000 feet, exploded over Walkersville, spreading a debris field of several square miles from Route 26 near Frederick to Woodsboro. Twenty-one people were killed, including the flight crew and their family members, and to this day the cause of the crash remains a mystery. This aerial shot shows amazingly little remaining of the aircraft that was the size of a Boeing 707. (Courtesy Susie Kyle.)

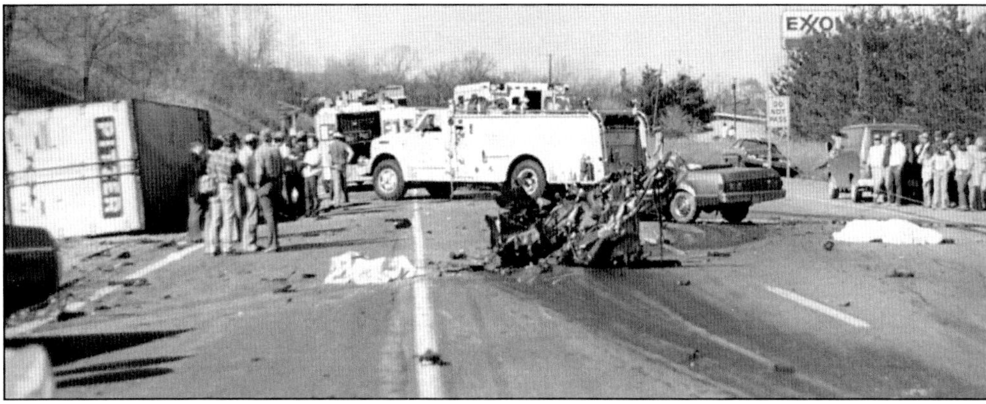

For years, U.S. Route 15 was considered the most dangerous stretch of highway in America. This fatal accident, near Ben's Exxon in Thurmont, was one of many that required the response of the Guardian Hose Company. The number of serous accidents decreased substantially when the road was converted to a dual-lane highway. (Courtesy Susie Kyle.)

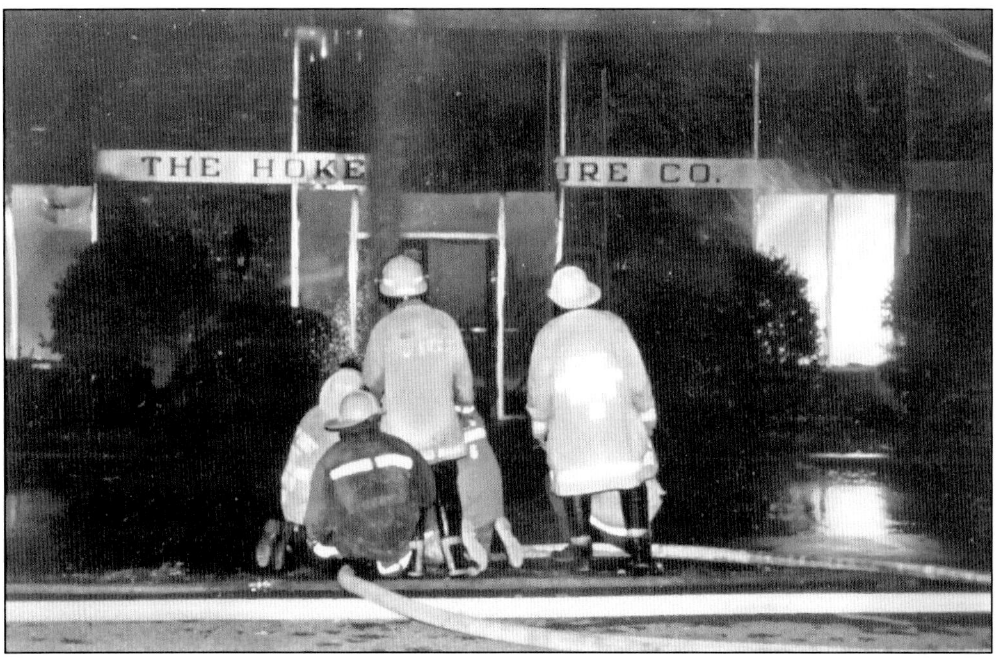

One of the largest fires in Thurmont history occurred in 1986, when the Hoke Furniture Company was completely destroyed. The fire, which destroyed the entire complex, required the help of over 100 firefighters from throughout Frederick County. (Courtesy RRFC.)

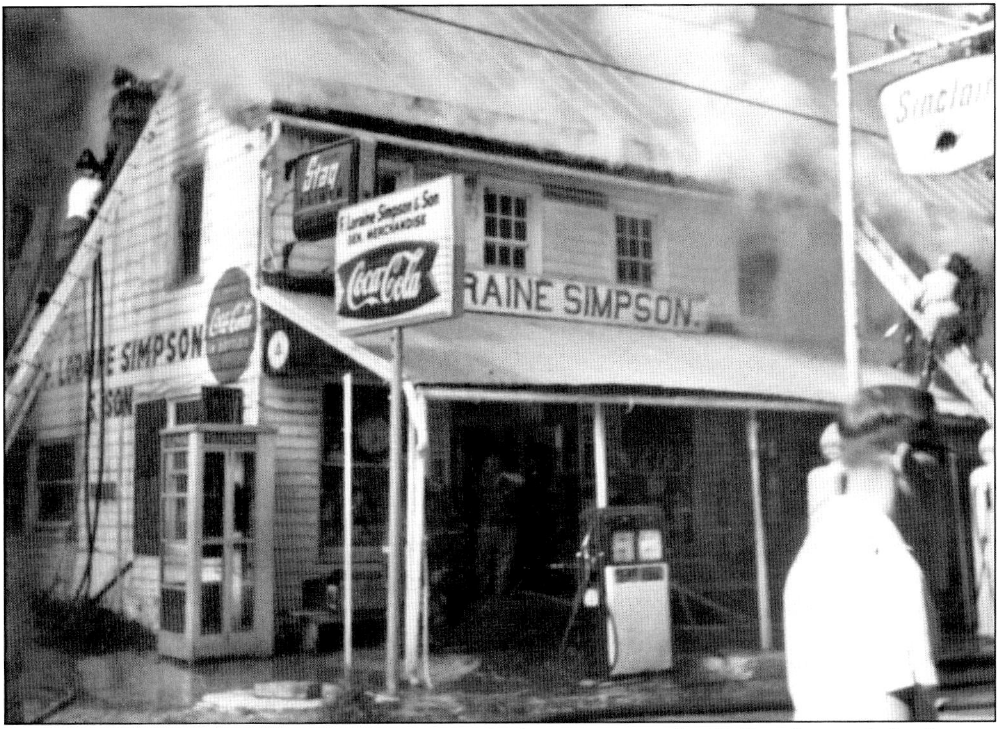

Two major fires spread through Main Street of Libertytown in the 1960s, ultimately leading to the organization of a fire company in 1964. The two fires occurred directly across the street from each other. The fire at Simpson's Store destroyed the store and spread to a nearby house, while the fire across the street destroyed three dwellings. (Courtesy LVFD.)

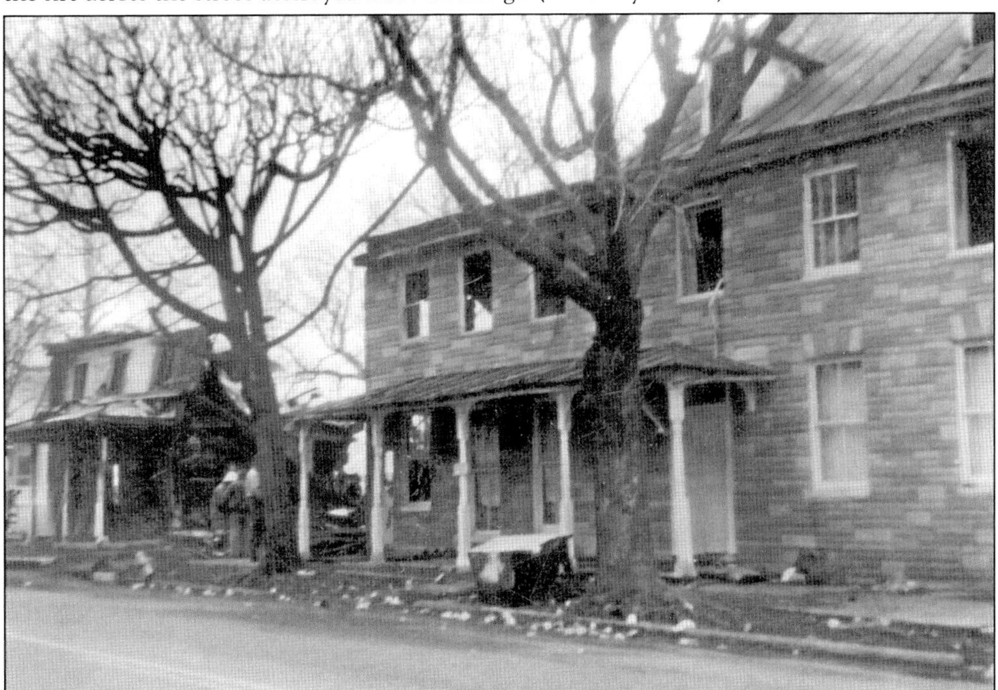

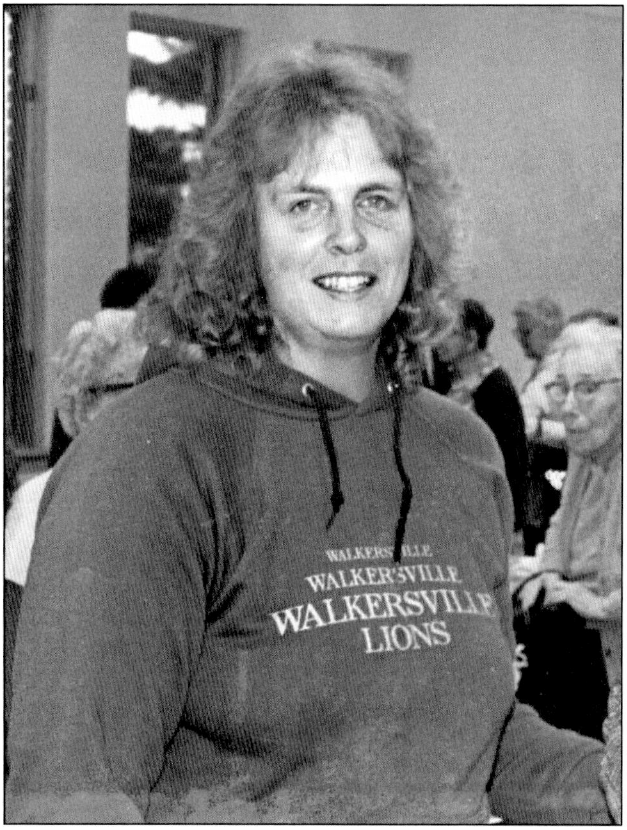

On April 2, 1987, Winona "Winnie" Crum responded with the Walkersville Volunteer Fire Company to a chimney fire on Old Frederick Road in Frederick County. While on the scene, another emergency call was dispatched nearby for a personal-injury auto accident. Firefighter Crum gave up her position in the cab of Walkersville Engine 114 to allow an EMT to respond to the scene. Firefighter Crum was riding on the rear step of the engine when, for unknown reasons, she fell from the apparatus and suffered fatal injuries. As of this printing, Firefighter Crum is the only female firefighter in Maryland to die in the line of duty responding to an emergency call. This book is dedicated to all those listed that have given their lives to serve the citizens of Frederick County. (Courtesy family of Winona Crum.)

Frederick County Deaths in the Line of Duty

Name	Company	Date
William Carlton	Junior Fire Company No. 2	December 22, 1840
Harry C. Root	Guardian Hose Company	June 10, 1898
William B. Davis	Junior Fire Company No. 2	March 20, 1910
Stanley Damuth	Guardian Hose Company	July 24, 1928
Walter K. Shank	Citizens Truck Company	November 24, 1957
Eugene J. McKissick	Guardian Hose Company	December 4, 1967
Howard M. Stroup	Citizens Truck Company	October 27, 1968
John F. Harp	Myersville Volunteer Fire Company	March 28, 1969
Gaither F. Frushour	Myersville Volunteer Fire Company	March 28, 1969
Samuel Kidwiler	Brunswick Volunteer Fire Company	March 24, 1971
Douglas Finneyfrock	Guardian Hose Company	June 21, 1973
Kenneth H. Willard	Walkersville Volunteer Fire Company	August 22, 1977
David Taulton	Brunswick Volunteer Fire Company	December 12, 1982
Winona L Crum	Walkersville Volunteer Fire Company	April 2, 1987
James H. Stavely III	Braddock Heights Volunteer Fire Company	June 30, 1992
Michael Wilcom	New Market District Volunteer Fire Company	May 24, 1993
Terry L. Myers	Vigilant Hose Company	February 15, 1999
Gilmore L. Stitley	Citizens Truck Company	January 10, 2000
Dora J. Wilt	Urbana Volunteer Fire and Rescue	September 13, 2001
Andrew J. Waybright	Frederick County Division Fire/Rescue Services (DFRS)	July 3, 2002